DOROTHY WORDSWORTH

Dorothy Wordsworth

ROBERT GITTINGS
AND
JO MANTON

CLARENDON PRESS · OXFORD
1985

Oxford University Press, Walton Street, Oxford OX2 6DP

London New York Toronto
Delhi Bombay Calcutta Madras Karachi
Kuala Lumpur Singapore Hong Kong Tokyo
Nairobi Dar es Salaam Cape Town
Melbourne Auckland

and associated companies in
Beirut Berlin Ibadan Mexico City Nicosia

Oxford is a trade mark of Oxford University Press

© Robert Gittings and Jo Manton 1985

All rights reserved. No part of this publication may be reproduced,
stored in a retrieval system, or transmitted, in any form or by any means,
electronic, mechanical, photocopying, recording, or otherwise, without
the prior permission of Oxford University Press

British Library Cataloguing in Publication Data
Gittings, Robert
Dorothy Wordsworth.
1. Wordsworth, Dorothy—Biography   2. Authors,
English—19th century—Biography
I. Title II. Manton, Jo
828'.709 PR5849
ISBN 0-19-818519-7

Set by DMB (Typesetting)
and printed in Great Britain
at the University Press, Oxford
by David Stanford
Printer to the University

To John Bell

Acknowledgements

The last scholarly biography of Dorothy Wordsworth, written by Ernest de Selincourt, appeared over fifty years ago. Despite the great merits of this work, a new Life must be needed to take into account much that has emerged in the past five decades, and, in particular, more recent editing and scholarship.

Our own book would have been impossible to undertake and to organize without the help of Professor Alan G. Hill, general editor of the newly-revised and expanded *Letters of William and Dorothy Wordsworth*. His assistance to us goes far beyond the fact that this edition has been our main source in portraying Dorothy Wordsworth's life. His interest in, and understanding of, our work, his patience with our problems, his tolerance of our ideas, and, above all, the friendly care with which he has answered our questions, have been exemplary. His kindness has extended to reading our typescript, and making creative suggestions for improvement and accuracy.

Among the many biographers of Dorothy's brother William, we have been happy in the kindly help of Dr. Mary Moorman, whose two-volume Life of the poet remains authoritative. Her assistance too has extended to answering specific questions with great patience and care, and giving us informally the benefit of her longstanding knowledge.

Among many individuals, who have placed special information at our disposal, we should like to thank Dr Roger Fiske, Alethea Hayter, Grevel Lindop, and John Woodforde. Special thanks must go to Dr I. I. J. M. Gibson, FRCP, Consultant Physician in Geriatric Medicine, Southern General Hospital, Glasgow, for her advice on Dorothy's complex medical history. She has let us use her observations as an Appendix, for which we are most grateful. Our special thanks go to Margaret Mullen for patiently typing our often-corrected manuscript.

Among those connected with organizations whose material we have used, we thank David Bromwich, Librarian of the Somerset Local History Library; Nigel Herring, Senior Keeper, Calderdale Museums Service; the late Peter Laver, former Librarian of the Wordsworth Library, Grasmere, and his staff; Dr T. T. McCormick, Director of Research, Dove Cottage,

Norma Watt of the Castle Museum, Norwich; Johnathan Wordsworth; and Dr Robert Woof.

The quotations on pages 105 and 106 of the text are printed by permission of the Houghton Library.

Acknowledgements for permission to reproduce the following, as illustrations, are due to Calderdale Museums Service for Plate 1; Norfolk Museums Service for Plate 2; the British Library for Plates 3, 8, and 9; the Somerset Local History Library for Plate 4; the Victoria and Albert Museum (Crown Copyright) for Plates 5 and 6; the Trustees of Dove Cottage for Plates 7 and 10; Richard Wordsworth and Mrs Mary Henderson for Plate 11; and the Abbot Hall Art Gallery, Kendal, for Plate 12. The silhouette of Dorothy Wordsworth on page 48 is reproduced by permission of the Trustees of Dove Cottage.

ROBERT GITTINGS *and*
Chichester JO MANTON

Contents

List of Plates

1. *'Dear Aunt'*

Dorothy Wordsworth first appears in Cumberland

at Cockermouth in the house where my Brothers and I were born and where our Father died. . . . It is at the outskirts of the Town, the garden bordering on the River Derwent or rather a *Terrace* which overlooks the River, a spot which I remember as vividly as if I had been there but the other day, though I have never seen it in its neatness, as my Father and Mother used to keep it, since I was just six years old.[1]

The setting of these brief six years, remembered so vividly, as Dorothy hints, because of their early loss, remained with her unchanged by time. Her ideal house, in later life, would have a terrace with a view. The river Derwent, broadening on its way to the sea eight miles away at Workington, still had the bright blue of a mountain stream. The sandy gravel of the water meadows blazed with yellow acres of groundsel and ragwort. The terrace, built to stop flooding, was a child's look-out battlement, while, safe in the house, the river's calm murmur was a reassuring, steady companion.[2] It mingled, morning and evening, with the voices from the secret nests of the birds in the deep-set hedge, privet and roses intermingled, which ran along the top of the terrace higher than its stone parapet. The tight weaving of the shrubs occasionally revealed the azure flash of a sparrow's eggs. The river banks were a tangle of mint and dock, feeding-ground of the woolly-bear caterpillars—'woolly boys' Dorothy called them[3]—later to emerge as scarlet tiger moths and pure white ermines.

The family house itself fronted the main street of Cockermouth, the most impressive private building there, with portico approached by stone steps, and eighteenth-century façade of many windows. It needed to be a large house. As well as Dorothy, born on Christmas Day 1771, there were two elder brothers, Richard (born 1768) and William (born 1770), and two younger, John and Christopher, born in 1772 and 1774, to say nothing of servants, a man in livery and a female staff, including a dependable housekeeper and a nurse for the growing brood of Wordsworths.[4] The Wordsworths were a Yorkshire family. The children's grandfather, Richard, an attorney, had

left his home, south of Leeds, earlier in the century, to work in
Appleby, on the northern tip of Westmorland, where he married
the mayor's daughter, Mary Robinson, and acquired a small
property at Sockbridge, near the neighbouring town of Penrith
in Cumberland. His son John, Dorothy's father, was born at
Sockbridge, and also became an attorney. He married Ann
Cookson of Penrith, the eighteen-year-old daughter of a linen-
draper, in 1766, having landed, just over a year earlier, an
important job, the post of law-agent to Lord Lowther. This
brought him to the big house at Cockermouth, further west,
which was a Lowther property, where Ann Wordsworth brought
up their children. She was remembered by them as a quiet,
peaceful, unobtrusive influence, yet nevertheless 'the heart and
hinge of all our learnings and our loves'.[5] 'I know', wrote Dorothy, 'that I received much good that I can trace back to her',
but unfortunately the brief remarks she hoped to make about
her mother do not seem to have survived.[6]

Little has survived, either, about the character of Dorothy's
father, apart from one laconic joke[7] he is said to have made
about his third son. Dorothy's father, John Wordsworth, was
employed for just under twenty years by the great Cumberland
landowner James Lowther. This was fortunate for him, though
it proved unfortunate for his children. John Wordsworth had,
through this employment, the benefit of a secure job and salary,
the rent-free possession of the handsome house at Cockermouth,
and a position of some influence in the district. On the other
hand, the reputation and dealings of his employer, in which he
had to take an intimate part, cannot have made him popular
locally, and this unpopularity woundingly fell upon John Wordsworth's own children. James Lowther, created Earl of Lonsdale
just after John Wordsworth's death, was one of the best-hated
men in the district, known throughout Cumberland and Westmorland as 'the bad Earl'; he was described as 'more detested
than any man alive . . . truly a madman although too rich to be
confined'. These riches put him above the law, particularly in
politics. He 'owned' nine Parliamentary seats, seven locally and
two as far afield as Haslemere in Surrey. His nominees for these
seats, when elected to Parliament, became known as 'Sir James's
ninepins' and had to vote precisely according to his orders.

John Wordsworth, as part of his duties, had to pay out 'large

sums' to freeholders at Parliamentary elections on James
Lowther's account. Often these were advanced out of his own
pocket. When their father died, nearly twenty years before his
patron, his children suffered from another trait of James
Lowther. He would not pay his debts, even to such a zealous
servant. Moreover, John Wordsworth had incurred such odium
by working for such a master, that others refused to acknowledge
debts to him. His daughter Dorothy, when fifteen, indignantly
instanced 'a Gentleman of my father's intimate acquaintance'
who would not pay a debt of £7,000 to the Wordsworth children
without 'considerable deductions'.

Dorothy's instinctive feelings were quick and impulsive from
a very early age. Her brother William, in often-quoted verses,
noticed how tenderly she felt toward the butterflies he unthink-
ingly pursued, and feared to brush the scales from their delicate
wings. He also remembered, all his life, the effect on her of her
first sight of the sea. This was not at the outlet of the Derwent
at Workington, but in the more dramatic setting of the port of
Whitehaven, a few miles south, where they had gone to stay with
their uncle, Richard Wordsworth. The children approached
'from the top of high ground down which the road descended
abruptly'; at the sight of the waves at the foot of the huge cliffs
and headlands, Dorothy, as William remembered, at once burst
into tears, 'indicating the sensibility for which she was so
remarkable'.[8]

These sensitive emotions were put to the extreme test, not
long after her sixth birthday. Early in 1778, her mother went to
visit friends in London, leaving the five children not at Cockermouth but with her own parents, the Cooksons of Penrith. Ann
Wordsworth returned to Penrith extremely ill. William afterwards hinted that she had suffered from some negligence.[9] The
family tradition was that she had been allowed to sleep in damp
sheets. Whatever the cause, she probably had pneumonia,
though the term 'a decline' generally indicated tuberculosis.

As she lay dying, Ann Wordsworth begged Elizabeth Threlkeld, the cousin who had been a witness at her marriage, to
make a home for her small daughter. She was buried on 11
March 1778 at Penrith; she was only thirty years old, and her
friend Mary Hutchinson came back from the cold funeral to sit
by the kitchen fire, warming herself and weeping at the

thought. For the five Wordsworth children under ten, it seemed
the end of their world. Yet thanks to her mother's forethought,
Dorothy Wordsworth had a happy childhood, for in June, faithfully as she promised, cousin Elizabeth Threlkeld appeared,
accompanied by her own brother. On 13 June 1778, Grandfather Cookson paid five guineas for places in a chaise to the
Old Cock Inn at Halifax, and she probably received ten guineas
a year for Dorothy's expenses. [10] Whatever the business arrangements, she gave the little girl a loving home.

This second cousin, whom Dorothy always called 'Aunt',
was then thirty-three. She came on her Cookson mother's side
from a family of Congregational ministers, and her father
Samuel Threlkeld had been for twelve years minister at Penrith.
In 1744 he moved to the Northgate End Chapel at Halifax,
where Elizabeth was born. She told the story of this journey on
horseback 'her mother behind a friend, her sister before the
man and her two brothers in panniers, out each side a horse.'
There was a drum or something hanging from the saddle. When
they got near Skipton, someone asked if they were players.
'Yes' said the friend, 'we shall play *Cato* tonight.' [11] This family
story later found its way into Wordsworth's *The Excursion*. Elizabeth Threlkeld grew up a lively girl, at Halifax, but deep-
rooted family loyalty kept her in close touch with her Penrith
cousins. Time would show what an excellent choice of foster-
mother Ann Wordsworth had made.

At Halifax, Dorothy Wordsworth found herself part of a large
family. Her 'aunt's' brother William Threlkeld lived nearby
with wife and children. Her new home was John Ferguson's
haberdasher's shop and house in Southgate where, since her
older sister's death, Elizabeth Threlkeld had taken care of
Samuel, Martha, Edward, Anne, and Elizabeth Ferguson, now
aged ten to seventeen. When her brother-in-law John Ferguson
also died, she took over the management of his haberdashery
business as well. [12] To these duties she added the care of six-
year-old Dorothy Wordsworth, with cheerful, capable affection, and Dorothy spoke of the mixed family as 'all her children'.
As the youngest, she was a favourite, and the cousins in later
life always demanded news of her, but at present they were too
old to be everyday playmates. Luckily she did not have far to
look for company. On the opposite side of Southgate lived a

flourishing wool-merchant William Pollard with his wife and a family which eventually numbered six daughters; one of these, Jane Pollard, was exactly Dorothy's age. From playing together, the two little girls grew up to be lifelong friends.

Dorothy ran in and out of the Pollards' house like one of their family, bursting into the parlour to hug them all and to join them round the dining-table, or in the happy circle of children surrounding Mrs Pollard's bright fire. 'Oh! how often has that Fire-place been surrounded by a party of happy children of whom I was not the least happy!' Jane watched from the window beside the same fireplace to see her friend cross the Back Lane and make for the door.[13] Once there, the whole house and its toys waited for them. Dorothy remembered 'the adventures of the baby-house, *the little parlour*, (I now fancy I see Harriot's shop fixed at one end of the long window-seat), the croft, the ware house, nay even the back kitchen'.[14]

As the two girls grew older, they began to move further afield. It was a welcoming countryside for their chief enjoyment, 'wandering wild together and shunning every other society'.[15] Halifax in the 1780s was still a country town; the first worsted mill was not opened until 1787 and the population at the 1801 census was still under nine thousand. Half a mile from the Pollards' house woodland covered a favourite valley where the girls roamed, black porringer in hand, picking bilberries. Each Midsummer Day was the Fair, 'to children the grandest day of the year, being always held on that day', when Dorothy 'used to hang out of the window by the hour to enjoy the Sports.'[16] Mossy paths followed the banks of the unpolluted river Calder, where a walker could

> catch the silvered glances of the trout
> Seen in the bottom of the lucid wave.[17]

The hills around were chequered with stone-walling, dividing the fields of independent farmer-weavers, each with his commoners' rights, cows, poultry, a house and in each house its spinning-wheel and tenter. For secrets or whispered confidences they took 'Mr. Caygille's walk', a lane running past his estate The Shayn up a steep wooded hill to the east of the town. Dorothy remembered how soundly she slept in childhood holidays after long days spent playing out of doors.

Even on Sundays Dorothy and Jane were not separated, for their two families, both Dissenters, attended the same meeting-house. Dorothy Wordsworth's education and upbringing were grounded in her aunt's religion, so this cannot be dismissed as a matter of small importance. Her aunt's father, as a Congregational minister, had undertaken to teach 'true religion and virtue' to the members of the Northgate End Chapel. The open trust of the Chapel excluded no one on doctrinal grounds, and after his death in 1767 the self-governing congregation 'with cheerfulness and unanimity' invited the Reverend John Ralph, whose teachings were decidedly Unitarian, to fill his place. He was the minister Dorothy knew as she went regularly with her aunt and the Ferguson family to what the locals called 'T' cellar hole chapel', because the congregation entered at gallery level and descended a flight of steps. Outside, Northgate End was a classical temple of 1762; the inside was a plain oak-panelled hall with box pews, the only decoration a 'velvet quishing', or cushion, for the pulpit. The services were simple; one lesson from the Bible, a sermon, metrical psalms 'lined out' by the clerk, and a prayer of intercession 'with much fulness' for particular persons, cases, and causes.

John Ralph taught his hearers the need for 'working out their own salvation' through character and responsibility towards others. He founded a library 'to promote religious knowledge . . . for the benefit of the poor'. Members of the chapel founded the Halifax Literary and Philosophical Society, the Mechanics' Institute and the Dispensary which became the town's infirmary. He spoke of the evils of 'excitement or exaggeration' in religion and deplored religious quarrels since 'ye wrath of man worketh not the righteousness of God'. Instead, he urged them to be open-minded; 'it behoveth us to try all things'.[18]

We have Dorothy's own word for it that these teachings appealed to her. The minister was one of the people she asked after when she left Halifax. 'I have a great regard for Mr. Ralph, so don't forget to tell me' she wrote.[19] Whatever the changes in her beliefs, the ethics of this upbringing stayed with her through life. From it came many of her most appealing characteristics: scrupulous honesty, a spirit of service, cheerfulness, above all the generous spirit which saw the best in ordinary men and women; 'how much more readily my heart receives the suggestions of Hope than of Despair'.

The Northgate End Chapel was a civilized and cultured society in this small provincial town. The Halifax physicians were traditionally members. The voluntary assistant minister 'beside being a rational Christian [he] was an excellent classical scholar'. Elizabeth Threlkeld's brother the Reverend Thomas Threlkeld, educated at the famous Warrington Dissenting Academy, was a meditative, myopic scholar, considered 'a prodigy of learning and of knowledge' who read nine or ten languages. Elizabeth herself was well-educated and capable, able to take on the management of her dead brother-in-law's business with calm cheerfulness. A girl educated in this society was sure to learn more than the traditional showy accomplishments of girls' boarding-schools. Dorothy herself frankly disclaimed these. For instance she was not 'taught to exercise the pencil'. 'My drawings will make you smile at my little skill', she wrote:[20] or, when a baby was delighted by her singing 'You will wonder at her taste'.[21] To Jane she wrote in 1793, 'You expect to find me an *accomplished* woman and I have no one acquirement to boast. I am still as was your old friend Dolly Wordsworth. . . . I have nothing to recommend me to your regard but a warm, honest and affectionate heart.'[22]

In 1781, at the age of nine, Dorothy, according to her father's accounts, was sent to a boarding-school at Hipperholme two miles from Halifax. This school, kept by Dr and Mrs Wilkinson, was small, co-educational and apparently happy. She remembered Dr Wilkinson and 'his wife, teacher of the Girls: and is a most excellent tempered, motherly and sensible woman as I know by experience'.[23] She only stayed there until 1784, however, for a reason not hard to guess. On 30 December 1783 John Wordsworth died. During the happy years at Halifax Dorothy had almost forgotten her father, since she never went home, even for Christmas, which was her own birthday. 'The day', she wrote rather sadly, 'was always kept by my Brothers with rejoicing in my Father's house, but for six years (the interval between my Mother's death and his) I was never once at home.'[24] Nor did she travel, through the bitter winter weather, to his funeral. 'It was a very snowy windy day on which my Father was buried and my three eldest brothers followed him to the grave; Christopher was at Penrith, and I was in Yorkshire.'[25] Yet if Dorothy felt little personal loss at the time, her father's death had immediate and lasting consequences for her.

Two uncles, Richard Wordsworth on their father's side and Christopher Cookson from their mother's family, became joint guardians of the orphaned children. Their duties were not enviable, for John Wordsworth had died intestate, his affairs in chaos. Lawyers are often the most dilatory people in anticipating their own legal arrangements. While advising their clients to look ahead, they very frequently fail to do so themselves. There was some excuse for John Wordsworth; no one would have thought that an active, energetic, capable man would die in his early forties, but he spent a shelterless winter night, lost on his way from work. His lack of provision had drastic effects for his children. He had a small personal estate, but nearly as much again, almost £5,000, was owed him for various transactions by his tight-fisted employer, James Lowther, Earl of Lonsdale. True to his local reputation, the Earl had no intention of paying up. The executor-uncles would not, naturally, send in a bill for the deceased's service rendered until it became clear that the Earl was not even going to pay a proportion of it. They did not therefore formally sue in the Court of King's Bench, for recovery of the whole, for another four years after Dorothy's father died. In actual fact, and in spite of complex legal action, the debt was never paid in the Earl's lifetime, a matter of twenty years after John Wordsworth's death. It was only paid then because the Earl's will fortunately stipulated that his just debts should be paid, and it was only then that the sum owing, which with twenty years' accumulated interest now amounted to no less than £8,500, began at last to reach the Wordsworth heirs. The uncle-executors, realizing from the start that they were in for a long legal process, took steps in anticipation of the Earl's non-payment. These meant certain economies involving the Wordsworth children, and Dorothy in particular, almost from the date of John Wordsworth's untimely death.

The first economy was to remove Dorothy from boarding-school. The four boys must, of course, receive a grammar-school education to equip them for professions, even if the guardians had to advance money of their own for school and college fees. For any girl this was, by the standards of the time, an impossibility. No professions existed for women, who were expected to marry, or failing this to make their homes, as Dorothy eventually did, with relations. For the moment, in 1784, her guardians paid a five-shillings entrance fee to another school in Halifax. This was not an unhappy move. Dorothy lived at home with her much-loved aunt, going each day with her friend Jane and the Pollard sisters to 'Miss Mellin's School'.[26]

Miss Martha Mellin and her sister Miss Hannah, were, like Aunt Threlkeld, attenders of the Northgate End Chapel; home and school influences were in harmony with each other. The academy later developed into a boarding-school in 'extensive premises near the town of Halifax', to which Dorothy's guardian Uncle Christopher brought his own daughter, so the education must have seemed to him satisfactory. Many of Dorothy's school-friends were the daughters of Halifax business- and tradesmen, and the lessons sensible, without frills. She stayed from the age of twelve until fifteen, a girl noted for lively cheerfulness, 'which seldom deserted me'. Three years after leaving school, she reassured Jane, 'my disposition is as cheerful, nay I think I may say as *lively* as ever'.[27]

A varied circle of school-friends survives in Dorothy's letters: 'my old companions whom I left mere girls', and still liked to picture to herself years later. There was Jane Pollard, with a round cap and childish ringlets on her shoulders. How hard to imagine Jane grown up! There was Patty Taylor, who made Jane jealous though 'she never held so large a share of my affection as you by many degrees'.[28] There were Jane's sisters, Ellen and Harriot Brachen, tall and fine-looking, Mary Grimshaw— 'rather fat . . . is she handsome?' Dorothy wondered when school-days were done. There were her own Ferguson cousins: 'I think I must be *nearly* as tall as Patty Ferguson'. Betsy Ferguson 'sadly deficient in taste, and used to curl her hair and put on her cap with less skill than any girl I know'.[29] Molly Waterhouse of Old Well Head by contrast was very pretty; moreover, she had 'Locke upon the Human Understanding, Euclid and several other such books' with her name in 'so she is you see quite a Learned Lady'.[30] Like all schoolgirls, they were interested in each other's weight. 'Do you grow fat? . . . I now reach *eight stone*,' enquired Dorothy proudly, for 'fat' was a fashionable term of praise.

Among these girls, with winter dances and summer picnics,

Dorothy Wordsworth grew up, receiving an unpretentious but solid education at home as well as school. Not even the progressive Dissenting academies were broadminded enough to consider admitting women students.[31] Yet women in Unitarian families gained much from the intellectual interests of their relations; reading aloud and conversation were in themselves an education. Mrs Barbauld, some of whose editions Dorothy Wordsworth later possessed, was educated in this manner. Philip Doddridge, a Dissenting minister whose works she also read, considered it of the highest importance to lead 'dawning minds into the knowledge and love of practical religion'. This demanded more than fashionable accomplishments, as the formidable Hannah More wrote in her *Strictures on Female Education*. A young woman should 'pursue every kind of study which will teach her to elicit truth, which will lead her to be intent on realities; will give prevision to her ideas; will make an exact mind'. Erasmus Darwin's *Plan for the Conduct of Female Education* stressed practical subjects to encourage 'health and agility of body with cheerfulness and agility of mind'. This was the tradition in which Elizabeth Threlkeld had been brought up and which she passed on to her much-loved youngest 'child'.

For, when due weight has been given to school, chapel, and friends, 'Aunt' was the centre of Dorothy Wordsworth's young life. She was 'my dear Aunt, my best friend'. For fifty years Dorothy continued to acknowledge her goodness. 'You know my heart too well to doubt my gratitude to her and affection for her', she wrote at eighteen.[32] Nearly twenty years later, at the age of sixty-five, 'Aunt' broke her hip, then usually a sentence to life as a cripple. 'What a blessing has her tranquil and cheerful mind been to her,' wrote Dorothy. 'She never speaks of her loss, but of the many enjoyments she possesses.'[33] Astoundingly for that date, by a system of exercises she learnt to walk again and continued an active life until the age of ninety-two. She dealt with illness briskly: 'half an ounce of Epsom salts in a tumbler of warmish water with a tablespoon of brandy' proved 'most efficacious'.[34] Dorothy thought her 'a *perfect* woman'. Her love for her children was not soft or sentimental; she could be angry at duty neglected and she expected from them the unselfish service to the family which she herself gave apparently 'all without effort, from a blessed nature'.[35] Dorothy's own life of

family devotion and delight in everyday things shows how well
she learnt from this deeply loved example.

  Dorothy Wordsworth became skilled and capable in all household matters, budgeting on a small income, accounts, cooking,
plain sewing, and general repairs. She wrote and spelled cor-
rectly, though she ruefully confessed to a grandniece, 'When I
was a little girl like you, I wrote very neatly and also what was
called a very good hand, but with making French exercises and
scribbling long letters to one of my companions'—Jane, of
course—'I fell into a careless way of making crooked lines and
irregular words and have never been able to get the better of
it!'[36] Above all she learnt an early and lifelong love of reading.
The Halifax Old Subscription Library, founded in 1769, was
housed for some years in her aunt's shop.[37] At fourteen she had
already rushed through the volumes of Samuel Richardson's
sophisticated *Clarissa*, which remained twenty years later her
standard of an enthralling book.[38] The society which allowed
a schoolgirl to read it was broadminded, and Dorothy Wordsworth, so pure in personal conduct, was not 'innocent' in the
conventional nineteenth-century sense of the word. Her taste
shows in the books her brothers gave her next year. 'I have a
very pretty little collection of books from my brothers, which
they have given me,' she wrote to Jane. 'I am determined to
read a great deal now both in French and English. . . . I am at
present reading the Iliad and like it very much.'[39] The collection
also included the *Odyssey*, Fielding's works, Hayley's poems, *Gil
Blas* in French, Goldsmith's poems, Milton's works and 'other
trifling things'. Brother Richard promised to send Shakespeare's
plays, which she wanted. She continued her education through
reading all her active life.

  Dorothy Wordsworth's mature character and intellect owed
everything to the happiness of these years at Halifax, above all
to her aunt's unwavering love. She knew this herself and summed it up in the simple statement 'you know she has been my
Mother'.[40] The love of these two was lifelong and for many
years Dorothy thought and wrote of her aunt's house as 'my
home'. The happiness, though, was not as secure as it had
seemed to a child. As she grew up, she found how much her
family's future lay in the power of the man she learnt in Whig-
gish Halifax to call 'Lord Lonsdale the greatest of Tyrants'.[41]

When her guardians eventually resolved to sue for moneys owed to the Wordsworth orphans, Lonsdale countered with an injunction in the Court of Exchequer. Legal wrangles followed. Later Dorothy was to report, 'Our affair with Lord Lonsdale goes on as usual which, I suppose, is very slowly.'[42] Until it was settled she could expect no security. At the age of fifteen, with a most painful shock, she learnt what this could mean. Before bringing an action, which must be long and hazardous, her guardians needed even further to economize on the expenses of their young ward.

2. 'Poor Dolly'

In May 1787 Dorothy Wordsworth's life changed suddenly, much against her will. She was summoned away from 'my dear Aunt, my best friend', from cousins, school, and childhood home to live with her Cookson grandparents over their draper's shop at the north end of the wind-swept market square in Penrith. To her they were strangers. Her grandfather, ill, had been she wrote in January 1788, 'these two years a burthen to himself and friends',[1] who 'never speaks to us but when he scolds, which is not seldom'. Her grandmother was harassed by the care of an ailing husband, and the shop where she sat every afternoon. Her interests lay in housework, the neighbours, and whether their servants were good workers. Moreover, as a staunch believer in original sin, Mrs Cookson was shocked by the hopeful views of nature and society in which Dorothy had been reared by Elizabeth Threlkeld; she found her granddaughter 'intractable and wild'. There was no money to keep the girl in idleness; she must learn to earn her living as other dependents did, by helping the relatives who gave her board and houseroom. Schooldays over,[2] there was no more time for dances and picnics, rambles over the hills, or talk of books. Dorothy sat silently mending an old shirt, while conversation, she wrote in August, was 'about *work, work*', sewing, and how 'such an one is a very *sedate, clever, notable* girl', inevitably praising someone she particularly disliked.[3] Penrith itself, a remote market town, after liberal, lively Halifax, she considered 'a petty place', riddled with gossip.

Bitterly conscious, as she was, of being a poor relation, her days were filled with domestic duties she secretly despised: 'Yet I am obliged to set upon the occasion as *notable* a face as if I was delighted with it and that nothing could be more agreeable to me.' Considering how tirelessly Dorothy Wordsworth worked all her life for people she loved, it is clear that her mother's family failed to win her affection at this time. Later, she was to write that both her grandmother and her guardian Uncle Christopher had been generous and kind to her in their own way.[4] Now she felt 'Never, till I came to Penrith did I feel the loss

sustained when I was deprived of a Father'. Not at his death
but now, she felt fully alone.

She was, after all, very young. Homesickness coloured all
her thoughts. 'I would give anything to go to Halifax instead,
to that dear place which I shall ever consider as my home. . . .
the loss of a Mother can only be made up by such a friend as
my dear Aunt.'[5] The house in the Southgate haunted her as
'the place where I passed the happy hours of my childhood'.
She waited impatiently for the return of her brothers from
Hawkshead Grammar School, blaming her uncle, Christopher
Cookson, that horses were not sent for them on the day term
ended—'indeed, nobody but myself expressed one wish to see
them'. Eventually William hired a horse and rode over himself
'because he thought some one must be ill'. When John and
Christopher followed, the orphans' conversations ended 'with
wishing we had a father and a home'. They realized now 'the
loss we sustained when we were deprived of our parents', and
'many a time have Wm, J., C. and myself shed tears together,
tears of the bitterest sorrow'.[6]

During these holidays Dorothy came to know three of her
brothers for the first time as individuals: William and Christopher clever, 'at least so they appear to the partial eyes of a
sister', and silent, shy John. 'You know not how happy I am in
their company . . . they are just the boys I could wish them,'
she wrote in July 1787; 'they are so affectionate and so kind to me
as makes me love them more every day.'[7] She dreaded the end
of the holidays and counted the hours anxiously. 'I have so few,
so very few to pass with my Brothers . . . Oh Jane! when they
have left me I shall be quite unhappy.' On 5 August the three
youngest returned to Hawkshead, John and Christopher for the
Autumn Half, and William to lodge until he went up to St.
John's, Cambridge in October. In the next year 'our poor John,
called a dunce',[8] was to go to sea, 'much delighted with the
profession he has chosen', in the East India Company's fleet.
Richard, a shrewd, slow-moving young man, 'diligent and far
from being dull', was already an articled clerk to his lawyer
cousin Richard Wordsworth of Braithwaite. It seemed that
only in her heart could she hold these scattered brothers
together as a true family.

On the day after the boys left, Dorothy sat up until eleven,

pouring out her feelings to Jane. 'I am quite alone. Imagine me
sitting in my bed-gown, my hair out of curl and hanging about
my face, with a small candle beside me and my whole person
the picture of poverty . . . and you will then see your old friend
Dorothy.' The days were forlorn and dull; her grandmother
seemed cold and unaffectionate; 'while I am in her house I cannot at all consider myself as at home; I feel like a stranger. You
cannot think how gravely and silently I sit with her and my
Gfr. . . . I sit for whole hours without saying anything.'9 William returned for three weeks in October, but she had little free
time to spend with him, as 'I was very busy during his stay preparing his cloathes for Cambridge'. He left, the northern winter
closed in, and she was alone.

Dorothy was far from unique in her troubles. The broken
home was as much a commonplace of the nineteenth century as
the twentieth, usually, as for the small Wordsworths, through
the death of a mother in illness or childbed. If the father did not
remarry swiftly, the standard solution to the problem of mother-
less children was to distribute them among various relations,
with whatever allowances the family could afford. Often, as for
Dorothy with Aunt Threlkeld and her lively household of children, this was a happy arrangement for everyone concerned.
In other ways though, Dorothy was unfortunate. A passionately
affectionate temperament made her vulnerable to the pain of
parting. The failure of her father to have her home for even the
briefest visit remained an obscurely hurtful mystery, even in
adult life. 'I was never once at home,' she wrote on her thirty-
fourth birthday' 'never was for a single moment under my
Father's roof after her death, which I cannot think of without
regret.'10 More serious even than the loss of her first family
home was that of the second, where she was growing up loving
and loved, happy and as she believed secure. The shock of
removal at the age of fifteen, for reasons she could not fully
understand, to the odious position of poor relation among
strangers, affected her profoundly. She remained until old age
proudly independent, even when almost destitute, saying 'I
fear not poverty'11 and rejecting offers of help. Indifference to
material wants was more than balanced, though, by the urge to
find a family and home where she belonged by right. All she
had lost, home, parents, and emotional security, was vested in

the four brothers who were all in all to her. 'Neither Absence
nor Distance nor Time can ever break the Chain.' While they
lived she told Jane, 'I shall never want a Friend.'[12] Family and
home, 'a blessing which I so early lost',[13] became as essential
as the air she breathed. She never lost this need.

Her way of expressing it, in letters and journals, is couched in
language which, to modern ears, may sound exaggerated, but
which was part of the common literary vocabulary of her upbringing, 'the vocabulary of sensibility', as it has been called.[14]
Both Dorothy and her friend Jane wrote in it, and continued to
do so, to some extent, all their lives. To realize this may save
one any surprise at some of the later expressions Dorothy uses.
When she calls her brother William 'my Beloved' or her infant
nephew 'noble', when, as now, she writes to Jane, 'how dearly
do I love you! no words can paint my affection', when, later
and with another friend, she dares not open a letter, 'afraid to
look into the terrible history of what you had endured', her expressions are not insincere or exaggerated, nor do they imply
relationships which modern usage might suggest. She is simply
using the vocabulary in which her tastes were formed. At this
time in Penrith, particularly to Jane, she is writing largely in
the style of Samuel Richardson's *Clarissa*, whose thousands of
pages the two girls had just shared. With her swift emotional
response to any situation, real of fictional, she felt herself the
unhappy heroine of that great novel, written totally in the form
of letters, such as she was writing to Jane. Their relationship
was to be exactly the same as that of the two girls in the first
part of *Clarissa*, where the heroine, forced to lead a restricted
life by her family, finds her only confidante in a girl of her own
age, Anne Howe. Dorothy's correspondence here is based on the
language of her favourite novel. The terms of endearment are
practically the same. 'I often wish', wrote Dorothy, 'that you
my Dear Jane were my sister, I think how happy we should be!'[15]
'Could you have been my sister!' Clarissa writes to Anne.
'There are souls', writes Clarissa, 'that can carry their friendships beyond accidents.' 'I am sure' Dorothy later promised
Jane, 'that love will never find me closer to any human Being'.[16]
It was easy for Dorothy to see the commonplace relatives at
Penrith as the family of Clarissa, who endures ill-nature not
only from mother and father, but from gossiping servants. She

vividly pictures herself in the same situation, 'found fault with every hour of the day both by the servants and my grandfr and grandmr. . . . I have no doubt but that they always conclude their conversations with "they have nothing to be proud of".'

It was, in fact, at this time that the orphans' finances, obscure through the two and a half years since their father's death, became clear. 'We shall have, I believe, about six hundred pounds apiece. . . . It is but very little, but it will be quite enough for my Brothers' education.' Yet Dorothy does not seem to have understood the finances she hopefully proclaimed. The sum of £600 each was indeed 'but very little'. Put on deposit at the Carlisle Bank, which the Wordsworth family used, it would give at 4 per cent interest an income of £24 a year for each child. This was far from 'quite enough for my Brothers' education', and it is not even sure that Dorothy herself ever actually received this minimal annual sum. The usual minimum for living at that time was about £65 per annum. If special outlay was required—university education for William and, later, Christopher in this instance—money must be advanced. Uncle Christopher reported with alarm that 'very extravagant' William overspent £300 in two years, and he left Cambridge owing £400. 'My lordly dressing-gown, I pass it by,' he wrote jokingly; but it was no joke to his uncle-trustee. As for poor Dorothy, £24 a year would barely cover between half and a third of her keep. It was not until eight years later that a fortuitous windfall enabled her and William to support themselves, without drawing on the charity of relatives. The expense of travel was probably the reason Dorothy was not allowed to accept an invitation to spend Christmas 1787 with her other guardian, Uncle Richard Wordsworth at Whitehaven, a bitter disappointment, since John and Kit would be there.[17] The likelihood is that she never received what was nominally due to her. Asking for her 'allowance', in a time of crisis in July 1793, she received what afterwards proved to be only a loan.[18]

Yet in spite of poverty, Dorothy's letters to Jane record normal pleasures, including the company of new friends of her own age in Penrith itself, girls with whom she could share many delightful experiences. These were the Hutchinsons, Mary and Peggy, orphans like herself, living at Penrith with their aunt

Elizabeth Monkhouse, and their severe great-aunt Gamage, a disciple of the Congregationalist Dr Watts. Dorothy's new friends shared her position as, she thought, 'at the mercy of ill-natured and illiberal relations'. Dorothy later wrote how

Mary and her sister Margaret and I used to steal to each other's houses, and when we had had our *talk* over the kitchen fire, to delay the moment of parting, paced up one street and down another by moon or starlight.[19]

In a small town like Penrith, half the size of Halifax, it was possible to know virtually everyone. The miniature market town largely consisted of one long street, running north and south. As a contemporary guide-book put it, 'There are many modern and well-built houses, in good taste here, and the inhabitants are wealthy, courteous, and well-bred.' The only eyesore was the slate-roofed shambles in the market-place, but from higher ground there were pleasant views, wherever the eye might look, over 'as rich meadows as any in the north of England'. From the look-out Beacon, a square building in the local red sandstone, set on the heights above the town and commanding a view of a hundred miles, one could see as far as the Scottish border to the north, and the crouching mountain mass of Skiddaw to the west.[20] South and near at hand one would see Brougham Castle and the spreading Lowther Woods, a favourite Sunday evening walk for the three girls. They would enter the woods below the mill, and walk along the path beneath the trees with the light from the west shining richly through the leaves, as far as the stone-quarry—'the boundary', Dorothy later wrote, 'of some of the happiest of the walks of my youth'.[21] The three girls made up a group of strong contrasts. Dorothy was active, impulsive, short and slight, only five feet high, the diminutive 'Dolly' as she called herself. Mary, two years older, was tall, dark-haired, white-skinned. She had a slight cast on one eye, which, however, did not spoil the sweetness of her countenance, and she radiated calm. Peggy, deceptively bright-cheeked, was to die unexpectedly young. Linked in friendship, they formed a striking set of young women, and Dorothy's mind still dwelt on their affection at this time in Penrith for many years afterwards.

Although marked by early sorrows, Dorothy reveals herself, when her first homesickness faded and she found friends, as a thoroughly normal young woman. With Jane she exchanged opinions on hair-curling, hats, and high heels. 'So you have got high-heeled shoes [? I do not] think of having them yet a while I am so little and wish to appear as Girlish as possible. I wear my hair curled about my face in light curls friz'd at the bottom and turned at the ends. How have you yours? I have tied my black hat under the chin, as it looked shabby in its former state.'[22] In December 1787 she asked for news of the theatre: 'I understand the players are at Halifax, have you ever been? is there any talk of building a new house?'—the promised Halifax playhouse which eventually opened in 1790. She wishes Jane 'many merry evenings and agreeable dances', but reported scathingly on the social life of Penrith: 'the assemblies are indeed begun, but they are no amusement for me, there was one on Wednesday evening where there were a number of Ladies but alas! only six Gentlemen, so two ladies are obliged to dance together'.[23] A letter ends, 'I begin to find myself very sleepy and I have my hair to curl, so I must bid my very Dear Friend a good night.'[24]

It is hard to imagine in the writer of these artless confidences the author of Dorothy Wordsworth's letters and journals. Yet she was always intensely responsive to affectionate influences and a new stage in her education had already begun. In the lonely winter of 1787, a friend appeared. Her grandparents' unmarried son, the Reverend William Cookson, was living for a time in their house, and volunteered to tutor his intelligent, frustrated young niece. From nine until eleven every morning he taught her to read and write French, Arithmetic, with the promise of Geography to come: Classics, then the basis of any serious education, were generally agreed to be too difficult for women. Dorothy was lucky in her teacher. William Cookson had gone from the draper's shop in the square to Sedbergh School. Although the Cooksons were pillars of the Congregational community in their home village of Salkeld, he became a member of the Church of England, and therefore was able to qualify for nomination as sizar then rapidly as scholar to St. John's, Cambridge. At his BA degree in 1776, he emerged as Fifth Wrangler, and in 1778 was elected to a fellowship. In 1780 he had been tutor to the sons of George III,[25] and having taken orders, was now, at the age of thirty-three, awaiting a college living. This cultured, kindly man won Dorothy's heart, and

became 'that Uncle whom I so much love . . . every day gives
me new proofs of his affection and every day I like him better
than I did before'. Apart from his distinguished teaching, he
was understanding; seeing her 'pinched for time' by household
duties, he allowed her to sit beside him and write to Jane in
comfort. He was evidently a power in the land of his formidable mother for, 'I sit in his room', wrote Dorothy in November,
'where we have a fire',26 by contrast with the cold bedroom in
which she normally scribbled her letters. Uncle William even
offered hope of escape from hated Penrith. He took her with
him on his visits to the household of the Master of Penrith
Grammar School, the Reverend John Cowper and his good-
natured daughter, another Dorothy. 'I often go to Mr. Cowper's
and like Miss D.C. better than ever. I wish my Uncle and she
would but get married,' wrote Dorothy Wordsworth ingenuously. 'I must certainly pay them a visit.' This depended, of
course, upon his getting a parish, but she went on hoping he
was 'now in the road to preferment'.27

At Christmas 1787 Dorothy sent Jane a thimble in return for
a home-made pocket handkerchief, and enclosed a lock of hair
'to remind you of poor Dolly whenever you see it'. More
serious was the death of Grandfather Cookson who had been
failing for two years. He was buried at Penrith three days
before Christmas, and in January 1788 'we still continue in the
same house and I believe . . . we shall continue here'. It seemed
nothing would change. There were, of course, enlivening inter-
ludes. Towards the end of his first Cambridge Long Vacation,
in September 1788, William joined her rambles with the Hutchinsons, and spent some time in the favourite walks upon Brougham Castle, the Penrith Beacon, and the neighbouring fells.
After over a year's separation, this seemed to brother and sister
'a gift then first bestowed.' He noticed, too, in their walks, the
quieter attractions of Mary Hutchinson.28

In October 1788, soon after William's return to Cambridge,
came the moment for which Dorothy had longed. On a walk at
Penrith, Uncle William and Miss Cowper told her a secret;
they were to be married in ten days' time and to move to his
new parish in Norfolk. Moreover, 'my happiness was very
unexpected', for they invited her to live with them. 'To live in
the country and with such kind friends! . . . I was almost mad

with joy; I cried and laughed alternately.'[29] The ten days passed
in a flurry of making and mending clothes. On 17 October
Dorothy was a witness at the wedding of her uncle and aunt,
the other witness being Mary Hutchinson's uncle. After a wedding breakfast at the bride's home they escaped the 'awful
forms' of reception, by setting out at once on their journey, the
first, though not the last, time Dorothy Wordsworth was to
form a third party on a honeymoon. They travelled by Newcastle, where they stayed a fortnight with relations, and by
Cambridge, where they saw William, 'very well and in excellent
spirits'. Cambridge delighted Dorothy. 'I could scarcely help
imagining myself in a different country when I was walking in
the college courts and groves; it looked so odd to see smart
powdered heads with black caps like helmets, only that they
have a square piece of wood at the top, and gowns, something
like those that clergymen wear; but I assure you (though a
description of the dress may sound very strange) it is exceedingly becoming; we only staid a day at Cambridge, as you may
be sure we were anxious to see our destined abode.'[30] Dorothy
did not say she would have liked to stay longer; William,
though loved like all her brothers, was not the centre of her
being he afterwards became. He may himself have seemed to
her somewhat 'odd' and 'smart' in his fresh guise, 'suiting
gentleman's array', as he himself wrote,

> In splendid clothes, with hose of silk, and hair
> Glittering like rimy trees . . .

and with his 'smooth housekeeping within' his rooms in the
First Court of St. John's College.[31] After four terms, he was
well embarked on his new life, while for the present his sixteen-
year-old sister's hopes were set upon her new life as the Rector's
niece, at Forncett St. Peter.

3. 'The Oeconomy of Charity'

The parish of Forncett St. Peter with Forncett St. Mary was scattered and included several hamlets. The advowson had been bought in 1725 by Dr Hill, whose will obliged his heirs always to present a Fellow of St. John's, Cambridge; it thus became a college living. Dr Zachary Brooks the previous rector died in August 1788 and by this comfortable arrangement William Cookson succeeded him.[1] Half-way between Norwich and Diss, among fields scattered with trees but then empty of houses, three winding lanes meet near the round flint tower with its five bells. Like many Norfolk churches, St. Peter's was icy; 'very small Congregation, it being so intensely cold,' wrote a local vicar on 4 January 1789.[2] Beside the church, a long gravelled walk under an avenue of lime trees leads to the Rectory beyond. Early in the eighteenth century the college had built a dignified red brick house with scrolled gable ends; a classical doorway admits the visitor to a square hall, with the Rector's panelled study on one side and a large family room on the other. The wooded garden slopes gently down to a small stream, and the loudest sound is still, as Dorothy Wordsworth heard it, the wind in the branches. She had come from one isolated corner of England to another, but, though there was little prospect of young company, she did not complain of loneliness.

Filled with gratitude and affection for her uncle and aunt, Dorothy entered whole-heartedly into the spirit of their rectory. As an undergraduate, and later a Fellow of St. John's College, Cambridge, William Cookson had formed a lasting friendship with a diminutive, devout contemporary, William Wilberforce. Through Wilberforce he was connected with the most famous of all Evangelical groups, the Clapham Sect. Evangelicals, clergy and laymen, rejected alike the frank worldliness of the Hanoverian Church and the callow frivolity of fashionable society. In place of brutal sports, gambling, and gold lace, they offered sobriety, black coats, the sanctified Sabbath. Above all, they preached the 'saving' of the individual soul, whatever its outward casing. They were therefore led to the needs of the poor, the ignorant, the exploited, and to the duties of practical Christianity. The spirit which inspired Wilberforce's Parlia-

1. 'that dear place': Halifax in 1775

2. 'quietly though very happily at Forncett': Forncett St Peter Church and Rectory

mentary campaign against the slave trade directed William
Cookson in the humdrum duties of his country parish.

Dorothy, now just seventeen, took her part in this work with
the imaginative sympathy in which she was so rich. 'We have
sketched out a plan of the manner in which we are to spend our
time', she told Jane in December 1788, making it clear that her
own wishes had been consulted. 'We are to have prayers at nine
o'clock (you will observe it is *winter*).' After a morning's reading and writing, 'We are to walk or visit our sick and poor
neighbours till three, which is our dining hour.'[3] Three weeks
later she was confident: 'We have, I think, visited most of the
poor people in the parish—my Uncle will I am sure do a great
deal of good in the place.'[4] In or about June 1789 Dorothy
took up another traditional duty of the vicarage lady, a Sunday
School. This was a serious undertaking, for country girls were
early pressed into service in kitchen and field, and this was the
only education many were likely to get. 'I have nine scholars,'
she wrote. 'Our hours in winter are, on Sunday mornings from
nine till church time: at noon from half past one till three: and
at night from four till half-past 5: those who live near us come
to me every Wednesday and Saturday evening.' Her teaching
plans were modest and entirely conventional. 'I only instruct
them in reading and spelling and they get off [that is, learn by
heart] prayers, hymns and catechisms. . . . We distribute
rewards such as books, caps, aprons & c.'[5]

The Cooksons and Dorothy had hopes of 'a school upon a
more extensive plan' with a mistress to teach the little girls
spinning and knitting on weekdays, while she continued the
Sunday reading lessons. 'We are going to establish a sort of
School of Industry; my Uncle is at present in Treaty about a
House for the purpose; my Aunt and I are to superintend the
Business.'[6] This was a typical venture of the new philanthropy
which united them, but the little school was liable to interrup-
tions, after Aunt Cookson gave birth in 1790 to Mary, 'the
sweetest and most entertaining child of her age I ever saw'.
The new mother feared a threat to her baby from 'the small
pox which have never been out of the parrish' in the intensely
crowded labourers' cottages, since 'we think my little girls likely
to bring them'. During the enforced holidays, whenever she
met one of the children, Dorothy was asked the 'flattering'

question, 'Pray Miss when shall we come to school again?' It is
clear that she did this work, not from a sense of duty or obliga-
tion to her uncle, but of her own free will, with pride and satisfaction. 'One of them who came to me six months on Sundays',
she reported to Jane, 'is able to read exceedingly well in the
'testament and can repeat the catechism and a part of an explanation of it; five or six hymns, the Lord's prayer, the creed and
a morning and evening prayer.' She concluded triumphantly
that the child 'did not know a letter when she came to me'.[7]

Meanwhile her good works received the highest seal of ap-
proval. In December 1789 William Wilberforce, one of the four
Members of Parliament for Yorkshire, came to stay for 'rather
better than a month' with his old college friend William Cookson. Dorothy was quickly fired with enthusiasm for the cause of
anti-slavery and hoped Jane's father would vote for Mr Wilberforce, 'one of the best of men'.[8] Wilberforce was comfortably
certain about the position of unmarried women; they 'may
always find an object in attending to the poor'. Moreover, he
was struck by the vitality of this young girl, her words tumbling
from her mouth with an almost stammering intensity. He gave
her ten guineas a year to distribute to the poor 'in what manner
I think best',[9] an astonishing opportunity for Dorothy, who
had been almost an object of charity herself. Spending the
money was more than a pious exercise, for the poverty was
real. Labourers who lost work and commoners' rights through
the enclosure of farm lands might be forced to re-enter the poor-
house three or four times in a year. At Gressenhall, fifteen miles
off, some hundred bastards under fourteen slept three to a flock
bed, 'as clean as can be expected'.[10]

Wilberforce also provided evangelical books, which Dorothy
Wordsworth dutifully read and tried to follow: 'a little treatise
on Regeneration', the New Testament with a commentary by
the Dissenting preacher Philip Doddridge, and for practical use
The Oeconomy of Charity by the tract-writer Mrs Trimmer. Sarah
Trimmer opened her first Sunday Schools at Brentford in 1786
and Queen Charlotte consulted her about establishing Sunday
Schools at Windsor. The *Oeconomy of Charity* was a guide for all
who wished to further this work. 'It is a most interesting employment to assist in instructing the poor children,' wrote Mrs
Trimmer. 'To see them hunger after spiritual food—who would

not exert his best endeavours.' This was the spirit in which Dorothy approached the work. From Mrs Trimmer, too, came her plans with her aunt for spinning and weaving lessons in a School of Industry.

It may seem inconsistent that Dorothy, brought up in the rational, progressive optimism of Halifax, should enter so swiftly into the soul-saving concerns of an evangelical rectory. Yet the inconsistency was itself consistent and lasted all her life. She had a generosity of feeling, which overflowed in ardent enthusiasm for the beliefs and concerns of people she loved. To support them was her chosen work and the writings we treasure were to her a by-product. 'What have I to communicate', she wrote to a correspondent, 'but our daily goings-on (which hardly vary from day to day) and my own peculiar feelings; and to make these interesting love must be in your heart.'[11] From this modest stance she never shifted.

So, she took up good works, and there was the more time for them as Forncett society proved limited. They drank tea with a widowed lady, given to 'censoriousness'; there was a 'respectable and worthy' clergyman in the next parish, and two goodtempered, unaffected Miss Borroughses with their mother, 'but unhappily her mind is frequently deranged'.[12] In spring 1791 Dorothy made three visits to Norfolk's social centre, the city of Norwich ten miles away, which yielded three visits to the Theatre Royal; 'but my principal errand was twice of a disagreeable nature; I went each time to get a tooth drawn',[13] which somewhat overshadowed the jollity. In any case, Dorothy protested to clearly unbelieving Jane, she did not enjoy '*routs* which are of all things in the world the most disagreeable'.[14] Going into company always gave her violent headaches; after her visit to Norwich, for which her grandmother gave her a new gown, she returned 'jaded and pale as Ashes' and she felt 'sure I should make the worst rake in the world'. While Jane enjoyed the diversions of a visit to Leeds, 'I have been sitting quietly though very happily at Forncett without having been at one Ball, one Play, one Concert'.[15] She joined in the plan of her aunt and uncle's days: household prayers, morning reading and writing, afternoon walks, three-o'clock dinner, 'and after tea my Uncle will sit with us', usually reading aloud in books borrowed from the Norwich Library, among them Hume's *History of England*.

The Cooksons were kind. 'My Aunt is without exception the best-tempered woman I know and is extremely kind to me.' They celebrated her birthdays and invited Aunt Threlkeld to stay in summer 1789 for the first meeting since Dorothy had left Halifax. She fell in with their conventional views apparently without question: a three-mile walk alone to the post at Long Stratton was adventurous, to sit on damp grass hazardous; Uncle William's connection with the Royal Family demanded household mourning even for the sinful Duke of Cumberland. '*We* are the only people to have taken his Royal Highness's death to heart,' she wrote in 1790. To observe court mourning was a sign of social standing. Hannah More was mortified to find herself the only person not in black at an assembly and Horace Walpole advised, 'it is much better on such an occasion to over than under do'. It is curious to find young Dorothy in their company. She waded through snow 'in half-boots and spatter-dashes', or walked to enjoy scenery, in Georgian terms, the country though not picturesque 'very pleasing, the surface is tolerably varied and we have great plenty of wood but a sad want of water'.[16] For private amusement she fed and tamed 'Robin redbreasts', until they would fly into her room, a pas-time that lasted well into her own very old age. Unlike Parson Woodforde's niece Nancy, who found Weston Longueville fifteen miles away, at the same date, dismally rural, Dorothy appeared perfectly contented.

In this quiet life every family event took on significance. The great excitement of autumn 1790 was to be told the secret of 'my aunt's' intended marriage. Elizabeth Threlkeld at forty-five accepted the proposal of her nephew Edward's employer William Rawson, a Halifax merchant, and married him on 7 March 1791 at Halifax Parish Church.[17]* This occupied all Dorothy's thoughts and, filled with joy for her loved foster-mother, she plied Jane with questions about the gown and the wedding. 'What sort of man is Mr. Rawson? . . . I wish to know whether he is grave or lively? younger than my Aunt or older? little or tall? fat or thin?'[18] Laughing at herself for 'this string of *womanish* queries', Dorothy still pressed for details of the wedding. 'I should like to know what visitors she had, how she was dressed & c & c & c. You will know how to interpret these & cs.' She

---

* Dissenting chapels were not yet licensed for marriages.

prayed ardently that God would give her aunt long years of happiness, 'my constant wish and prayer, as it must be of all her children'.

This delight in a wedding, even a middle-aged wedding, was most natural in a girl of eighteen. Yet for herself, Dorothy apparently demanded no such happiness. At the time of 'Aunt Rawson's' engagement, Jane teased Dorothy affectionately about a suitor for herself, suggesting most improbably the thirty-year-old William Wilberforce. Dorothy laughed at the whole idea; it was 'very improbable' that 'Mr. W. would think of me', he would look for 'a Lady possessed of many more accomplishments than I can boast'. More seriously she pointed out that 'no man I have seen has appeared to regard me with any degree of partiality; nor has anyone gained my affections, of this you need not doubt'. Her heart was 'perfectly disengaged'.[19] Jane, remembering her cheerful school-friend, treated this as a passing mood, and six months later returned to the ever-interesting topic of love and lovers. This time Dorothy's answer was short, and for her almost sharp. 'I cannot suppose you entertain any such improbable suspicions as you are pleased to hint. I shall think you unkind if you say anything more to me upon the subject.' Jane took this seriously and she was right to do so. The subject of love vanished from Dorothy's letters, to reappear only once, briefly, twelve years later when she declared, 'it would be absurd at my age (30 years)' to talk of marriage.[20] Beyond this there is nothing, in letters, journals, or poems.

It is sometimes said that Dorothy's intense love for William prevented the growth of love for any other man Yet this double declaration of sexual independence to Jane was made, and never retracted, exactly four years before brother and sister began their first weeks of life together. Dorothy, it seems, knew her own nature. Family life was all in all to her; she loved children intensely and took pride in domestic duties well done. She had a genius for friendship, entering with steady imaginative sympathy into the hearts of men and woman alike. Yet, with all her lovable qualities, there is no sign that she ever aroused or experienced physical desire, nor that she felt this a loss. An admiring and perceptive man noticed the 'unsexual' character of her body, gait, and manner, while delighting in her friendship. From girlhood she seemed destined to be a creature apart, one of that distinctive company of nineteenth-century woman, clinging like sterile buds to the family stem, tight-furled until November withered them.

Dorothy, at least in youth, delighted in marriage for others and there is evidence that she accepted without question the usual marriage arrangements of her time and class.[21] She rejoiced generously when in 1795 her dear Jane Pollard married John Marshall, a rich linen manufacturer of Leeds, and went off to the Lake District on their wedding journey. The two fathers were old friends, the young couple retired to a comfortable house, Jane in due course gave birth to twelve children, her husband became an MP, and all was as it should be.

Meanwhile, Uncle William Cookson provided for Dorothy as a member of his family, and for this she was grateful. She had a traditional part to play in the household at Forncett. As 'Cousin Dolly' she slept in a garret and helped in the care of the growing Cookson family. By the end of 1791 she wrote with eighteenth-century forthrightness that her Aunt expects 'another young one early in the spring. This is rather sooner than we could have wished'.[22] In April 'the birth of another little cousin has made me more necessary than ever to my Aunt'.[23] Between March 1790 and June 1793 Nancy, Christopher, William, and George were born. Dorothy 'made linnen' and 'put the house in order'; she carried the current baby in her arms, both indoors and out, and amused restless three-year-olds. She washed the children in the nursery and superintended 'the terrible operation of dressing'. She helped to nurse childish or infectious illnesses, no trivial matter then, and reported cheerfully in May 1792, 'our young ones have got charmingly through the smallpox'.[24] On a journey, her aunt confidently 'left the Children to my care and hastened forward'. At the birth of George in 1793 Dorothy was 'not in bed last night and was up at five o'clock the preceding morning'. She shared her aunt's bed to be near at hand, and cheerfully decided 'lying-in is not half so tremendous a Business as it is generally thought'.[25] A month later, she was still sleeping with her aunt and told Jane, 'As I am head nurse, housekeeper, tutoress of the little ones or rather superintendent of the nursery, I am at present a very busy woman and literally *steal* the moments which I employ in letter-writing.'[26]

All this was hard work, but good domestic experience, pre-figuring Dorothy's duties in the Wordsworth household in years to come. Any present-day reader might feel that her aunt and uncle exploited their impoverished young relation, but it is only fair to say that to them, as to her, by the standard of the times, the arrangement seemed entirely reasonable. She herself 'would do nothing inconsistent with the duty I owe my Uncle and Aunt'.[27]

A single woman, however intelligent and well-read, expected to share the domestic burdens of her family. Elizabeth Carter, the linguist, learnt Latin and Greek from her father, went on to teach herself something of Hebrew, French, Italian, German, Spanish, and even Arabic. She contributed to the *Gentleman's Magazine*. Her translation of Epictetus, when published, made her nearly a thousand pounds, and Southey carried it in his pocket for twelve years, 'till my very heart was engrained with it'; yet she wrote cheerfully of sewing her way through 'whole dozens of shirts and shifts', and invented the recipe for 'a special good sweet cake', which she baked for every family christen-ing.[28] Even keeping a girls' school was regarded by bookish Catherine Harrison's 'worthy aunts' in York as paid employ-ment liable to 'remove her from the rank of a gentlewoman'. To engage in trade, like Dorothy Wordsworth's grandmother, foster-mother, or in their turn her Ferguson female cousins, would have seemed to many genteel families beyond the pale. To be useful in one's family circle, by contrast, was seen as part of the natural order for an unmarried woman. By the standards of the time Dorothy Wordsworth's relations were not depriving her of any possible career, but rather providing her with the essential protection of a home. In return, Dorothy early schooled herself to a lifetime in the service of her family, not merely of necessity, but gladly dedicated to 'the idea of home'.[29]

The year 1792, Dorothy's twenty-first, brought a new experi-ence. On 20 January, her uncle was installed as Canon of Windsor, where he had been tutor to the princes, and planned to take his family for his three months in residence the follow-ing August. On 31 July the party set out, Dorothy travelling not in the family chaise but 'with two of the Man-servants in a Stage Coach'. She 'did not like London at all', except for the view from the top of St. Paul's. Hot weather and the care of the

children meant she 'could not even enjoy walking about', and
she was delighted to leave for Windsor, where, with the Royal
Family in residence, 'when I first set foot upon the Terrace . . .
I fancied myself treading upon Fairy Ground and that the gay
Company around me was brought there by Enchantment'.
Everything delighted her, the band playing, the well-dressed
crowd, the Queen and the dazzling Princesses, above all the
old King and his unaffected fondness for children. 'Mary he
considers as a great Beauty . . . the first time she appeared
before him she had an unbecoming and rather shabby hat on.
We had, then, got her a new one. "Ah", says he, "Mary, that's
a pretty hat"!' She watched Queen Charlotte, driving a phaeton
with four white ponies in the Little Park below the Terrace,
sometimes half-hidden, sometimes glimpsed through the trees;
'her Equipage and her Train of Horseman looked so diminutive
that it was impossible to avoid comparing them with the descriptions one has read of Fairies travelling on fairy Ground'.
Dorothy was convinced no one could see the Royal Family at
Windsor without loving them.*

   The social life was pleasantly intimate: 'we compose, as it
were, one large Family in the Cloisters, for we can visit each
other's houses without Hat or Cloak for they are surrounded
by a covered Passage as, I believe all Cloisters are'. They could
visit each other 'without Form which is the mode I like best'.
Dorothy went with two young ladies and an agreeable young
man, their cousin, for 'several very charming little Excursions
into the country', to Egham Races and to admire the view from
St. Leonard's Hill. Her aunt 'was so good as to give me leave
to go' to one of the monthly Balls. Dorothy at first refused, but
was persuaded to make 'my Entrée (for I was never in a public
Room before) at one of the Egham Race Balls', with her new
friends and with their aunt, Mrs Heberden, as the essential
chaperon. 'I had the most severe tremblings and palpitations
during the first Dance, that can be conceived by any trembling
Female. My Partner was a wretched one and I had not danced
for five years.' Once she danced with young Mr Heberden, 'with
whom I was very well acquainted I felt myself quite at Ease' and
the rest of the evening went off well. So ended Dorothy Words-

---

   * In the same month, William Wordsworth in Orléans exulted in the proclamation
of the French Republic.

worth's brief coming out. Yet the stay at Windsor had drawn from her the first of her spell-binding descriptive letters.[30]

There was a dark side to these, on the whole, contented years at Forncett from December 1788 until February 1794. All the time at work, walking, even among the splendours of Windsor, Dorothy missed her brothers. They were her own, bound to her by early loss, and no one could take their place. They haunted her in 'the shapeless wishes of my youth—wishes without hope'.[31] All her hopes or plans for life centred on them. Poverty held no terrors, 'for if we have sufficient to provide for my Brothers on them I know I may depend'.[32] Yet meanwhile the quotation from *The Merchant of Venice* haunted her: 'how we are squandered abroad'.[33] By October 1790, John, 'tall and handsome', had been on an eighteen-months' voyage to India in the *Earl of Abergavenny*. William was at Cambridge, Kit entered for Trinity College, and Richard, at Gray's Inn, coped with 'our vexational business with the tyrannical Lord Lowther' as best he could. The Cooksons were generous with invitations to stay at Forncett: William came briefly at the beginning of his second Long Vacation at the beginning of June 1789. In 1792 John stayed for four months between voyages, Richard visited after a case at the Cambridge Assizes, and Christopher, whom she had not seen for five years, spent his 1792 Christmas vacation at the Rectory. He proved 'a most amiable young Man, sensible, affectionate and engaging'.

These brief meetings, as Dorothy confessed to Jane, made her miss them all the more. 'I have passed one and twenty years of my life,' she wrote in February 1793, regretting 'that the first six years only of this Time was spent in the enjoyment of the same Pleasures that were enjoyed by my Brothers and that I was then too young to be sensible of the Blessing. We have been endeared to each other by early misfortune. We in the same moment lost a father, a mother, a home, we have been equally deprived of our patrimony by the cruel Hand of lordly Tyranny. These afflictions have all contributed to unite us closer by the bonds of affection, notwithstanding we have been compelled to spend our youth far asunder. . . . Neither absence nor Distance nor Time can ever break the Chain that links me to my Brothers.'[34]

All the same, Dorothy was happy at Forncett. The company

of her uncle and aunt, when the latter was not in childbed, was relaxed and agreeable. There was evidently time for little jokes of a familiar or topical nature. Dorothy had a quiet sense of humour, which spills over into her letters at this time. With Jane Pollard, she adopted the teasing attitude that any inconsistencies in Jane's own letters might be put down to the absentmindedness caused by some love-affair. She wrote to Jane about one such letter, 'As it had no Date (it was not without signature so I will only suppose you *half* in Love)'.[35] She would even joke about William's 'great attachment to poetry . . . which is not the most likely thing to produce his advancement in the world'.[36] When the welcome present of a beaver hat arrived from Richard by coach just too late for her to send her thanks immediately by the London mail, Dorothy makes a typical joke, familiar in the days of coaches and their coachmen, massively protected from the weather by their heavy double capes. She writes to Richard that one of John's seafaring mates should have been 'doubling the Cape'—i.e. driving the coach.[37] This was a familiar pun in the later decades of the eighteenth century, and was still going strong as late as 1819, when Keats used it, though he admits that by then it was pretty well worn out: 'perhaps', he writes of another coaching pun, 'that's as old as "doubling the cape"?' A girl with four brothers, Dorothy shows herself well up on the slang humour of the day.

Ever since Christmas 1790, when she was nineteen, her shadowy plans for the future had begun to concentrate on one of these much-loved brothers, whom she had not seen for eighteen months, and then only for a short stay at Forncett on his way north, at the beginning of his second Long Vacation. At Christmas and New Year 1790-1, her brother William arrived for a six-weeks' stay at the Rectory.[38] She saw a tall, gaunt young man, with high forehead, and Roman nose.[39] He was silent, even awkward, with strangers yet to his sister he spoke with 'a sort of violence of affection if I may so term it . . . a sort of restless watchfulness which I know not how to describe . . . a Tenderness that never sleeps'. Already William was in conflict with authority. Elder relatives had advised him 'to establish a reputation at College which will go with you and serve you thro' life'; yet from a passionate devotion to poetry he refused even to sit the mathematical papers which might have earned him a fellowship at St. John's.[40] His future was problematical.

All this Dorothy recorded clearly enough, yet William's presence promised magical possibilities. Although it was the depth of winter, every afternoon as soon as dinner was over they went out into the garden. Both of them particularly loved 'a moonlight or twilight walk' and they paced up and down the gravel path under the lime trees until the summons to tea at six o'clock.[41] The weather, at first very mild, grew colder, a blustery wind whistling through the trees above their heads, but Dorothy did not notice the cold. From this time onward she formed a firm resolve: somehow they would make a home together. Two years later it was still her dominant thought, though Brother Kit had shown himself almost equally attractive.[42] 'I am very sure', she wrote, 'that Love will never bind me closer to any human Being than Friendship binds me to . . . William, my earliest and my dearest Male Friend.'[43]

4. 'The character and virtues of my Brother'

There was only one stumbling-block to the scheme of a joint future, planned by Dorothy during Christmas 1790 at Forncett, and nurtured by her, with ever-increasing fervour, over the next two or more years. William's problems during these two years seemed to set the prospect further away. Shortly after his stay at Forncett, on 21 January 1791, he received his BA, without honours, at Cambridge. He did not, apparently, seek employment, but spent four months in London, living in lodgings on a remittance of £60 from his guardians—about enough to last a year. From the end of May, he spent nearly another four months at the home in Wales of his Cambridge friend, Robert Jones, with whom he had toured Switzerland the previous summer. Meanwhile, his relatives were anxious, and it was probably his uncle, William Cookson, who urged him to visit his influential cousin, John Robinson, MP for Harwich. Robinson saw him, perhaps at his house in Isleworth, and suggested a curacy at Harwich 'where his interest chiefly lies';[1] William countered by saying he was still eighteen months too young to take Holy Orders. Undeterred, Uncle William sent him to fill in time by being tutored in Oriental Languages at Cambridge;[2] but William stayed there only a few weeks until Full Term,[3] and did nothing about finding a tutor. By 6 November, he had contrived for the plan to change to his going to France, till next summer, to learn French, though he promised his uncle to take up the Oriental Languages directly he returned.

Leaving for France in the last week of November 1791, he did not, in fact, return to England for another thirteen months. In May 1792, he heard from Uncle William, still busy in his benefit, that he could himself make possible a curacy for William at Forncett, though to his own inconvenience. William's reply[4] seemed a little less than willing in tone. Dorothy, however, who also received a letter from him, was delighted, and indulged in her favourite pastime, described earlier to Jane as 'many different plans for building one Castle'.[5] She had a conversation with her uncle. Realizing it certainly would be inconvenient for him to keep the nomination of the Forncett curacy open for William, she thought it would be 'a charming thing' for William to be placed by Mr Robinson at Harwich, where the Vicar (Uncle William's own brother-in-law) had fled his parish to avoid creditors. There William could take the full duties of the parish; though 'he could only enjoy the profits as of a Curacy, yet it would in the End be a certain Provision'.[6]

None of this happend. William did not return to England for another seven months, in the second half of December 1792. When he did, he neither made enquiries at Forncett about the curacy, nor followed his promise to his uncle to seek a tutor in Oriental Languages. He simply stayed in London. The sole fruit of his past two years appeared to be two poems, published by the radical bookseller Joseph Johnson, advance copies of which reached Forncett some time in January.* Dorothy was there in company with her young brother Christopher, whom she had greeted with 'transports' after not meeting him for five years. Together they studied William's poems in a surprisingly critical manner, Dorothy objecting to the excessive use of unusually coined words, 'moveless' (three times) and, even more, 'viewless' (six times). Nevertheless, the poems in their conventional couplets, and particularly *An Evening Walk*, addressed to her, could not fail to cause her enthusiasm with their 'many Beauties, Beauties which could only have been created by the Imagination of a Poet'.[7] With equal enthusiasm she now painted to Jane Pollard all the details of the life she hoped to share with William in the Parsonage at Harwich, regardless of his hesitations about the clerical life. She wrote to Jane (16 February 1793):

I look forward with full confidence to the Happiness of receiving you in my little Parsonage. I hope you will spend at least a year with me. I have laid the particular scheme of happiness for each Season. When I think of Winter I hasten to furnish our little Parlour, I close the shutters, set out the Tea-table, brighten the Fire. When our Refreshment is ended I produce our Work, and William brings his book to our Table and contributes at once to our Instruction and amusement, and at Intervals we lay aside the Book and each hazard our observations upon what has been read without the fear of Ridicule or Censure.[8]

* The official publication date was the 29th, but copies seem to have come earlier.

If not 'one Castle', Dorothy certainly portrayed a very feasible Parsonage. This touching picture of all that Dorothy had worked for over the last two years was almost at once shattered by news in a letter from William in London.

She had been anxious about him in France as newspapers printed stories of war fever, King Louis XVI forced to declare war on Austria in April 1792, a Paris mob attacking the Tuileries in July, by August the royal family imprisoned in the Temple. At Forncett and Windsor, the doctrines of Liberty and Equality appeared 'new-fangled'. 'I begin to wish he was in England; he assures me however that he is perfectly safe.' Yet she felt, 'I cannot be quite easy.'[9] Her unease was justified, though for very different reasons. William had gone to Orléans in December 1791, according to brother Richard, as a 'less expensive and more improving' way of learning French. By 19 December, looking for lodgings, he had met a 'Family which I find very agreable' and had spent some evenings with them. 'I do not intend to take a master. I think I can do nearly as well without one,' he told Richard.[10] In fact he was being tutored in French by a charming girl.

Annette Vallon was the descendant of a family of surgeons in Blois, who held appointments at the eighteenth-century abbey building, between the church of St. Nicholas and the river Loire, which the Bishop had recently re-dedicated as a Hospital. On the far bank of the river, at the Church of St. Saturnin, her uncle was the parish priest. Annette's widowed mother had remarried and Annette was staying, strangely unchaperoned for a middle-class French girl, with her brother in Orléans. Both tutor and pupil were young, Annette twenty-five to William's twenty-one, both away from home, both open to the nameless exaltation of those who live through great events. In the dangerous, swiftly changing scenery of the Revolution, they shared a quickening of the senses, a sudden instant of passion.[11] Three months after their first meeting Annette returned to her family in Blois, already pregnant; in spring 1792 she was sewing clothes, including a pathetic 'petite toque rose', for William's child. While Dorothy was enjoying the decorous summer pleasures of the cloisters at Windsor, Annette was meeting William by stealth and trying to conceal her pregnancy. In September, when it could no longer be hidden, she retreated to

Orléans. William, lodging nearby, wrote urgently to Richard on 3 September 1792 for an extra twenty pounds. 'You will send me the money immediately.'[12]

William and Annette were separated by political opinions, she and her brother royalist to the point of danger, while he condemned the "antient guilty splendor" of the Catholic clergy and exulted on 20 September 1792 in the proclamation of the French Republic. More gravely, they were separated by language; Annette, so affectionate and warm-hearted, could never share her lover's lifelong dedication to the art of English poetry. Yet they were united by their first youthful passion and their unborn child. In this dilemma William went to Paris, where the guillotine was already at work and the National Convention on 1 December took the decision to try the king. For six weeks he read deeply in political literature and met at least two of the English radicals who had travelled to France in support of the Revolution. On 15 December, her father's name defeating the clerk, his and Annette's daughter, Anne-Caroline Wordswodst, born that same day, was baptized at the Church of Sainte Croix in Orléans. William was not present, but was legally represented at the ceremony by Annette's landlord. By 22 December, with feelings one can only imagine, William had returned to London. He wrote to Dorothy, who felt him 'a most affectionate brother',[13] and, while not at first revealing to her the personal dilemma of his life, continued to write regularly. On 21 January 1793 Louis XVI's head was cut off in Paris. Dorothy had felt it impossible to see the English Royal Family at Windsor 'without loving them'. In William's mind, the general English horror at the French king's death was merely 'the idle cry of modish lamentation', or so he wrote.

A greater test of family loyalty was to come. Soon after her happy and confident letter of 16 February inviting Jane to 'my little Parsonage', Dorothy learnt of the existence of Annette and Caroline. She responded without hesitation. France had declared war against England on 1 February, but the mail boats were still sailing. Impulsively she dashed off at least two generous and affectionate letters to her 'chère soeur'. She had always loved children; the Cooksons were for her 'our little People . . . the sweetest children I ever saw'. The fact that Caroline was illegitimate seemed not to shock her in the least.

Always more ready to hope than to despair, she spoke with enthusiasm of the 'petit ménage' which, regardless of war, they would somehow all share. Annette was bitterly unhappy, for when she returned with the baby to her home in Blois, her family had put Caroline out to nurse, and she suffered the grief of seeing her baby carried past the house like a stranger. Dorothy's generous love, her understanding, her refusal to condemn, 'cet intérêt si touchant que vous prenez à mes peines', comforted the poor girl. Perhaps after all William would return to marry her and 'ma fille aurait un père'. In fact with every day of war this grew less likely; mail boats ceased to ply in March and within six months any British man found in France was arrested. Yet it is striking how in this crisis Dorothy said nothing of her own disappointments or the threat to her long-cherished plans; her thoughts were all for the young mother and child. No wonder Annette hoped that Caroline might resemble her.

Meanwhile Dorothy faced an inevitable problem. Someone must break the unwelcome news of Caroline's existence to Uncle William Cookson. For a girl of twenty-one this was a frightening prospect. Annette wrote to William on 20 March 1793, 'l'engager de rien dire a ton oncle; ce sera un combat pénible qu'elle aura à soutenir. Mais tu le juge nécessaire.' This letter, like her reply to Dorothy, never reached England. Both were confiscated by the Blois police and survived, by historical irony, in the archives of the Département, where they emerged 130 years later to reveal Annette's story. William was still in London, staying with Richard who had just started his own law practice at 11 Staple Inn, Holborn. Revolutionary enthusiasts returned from France were highly suspect to the Home Office, and his presence was embarrassing to the cautious young lawyer, but Wordsworth blood was much thicker than water. Dorothy noticed there was no 'similarity of taste or sentiment' between the brothers, but just as she accepted William's opinions, so Richard gave him house-room.[14] At Forncett Dorothy was 'tourmentée', finding no one she could talk to freely and forced to hide her tears. In the end it is clear that the duty of telling Uncle William was left to her.[15]

She seems to have kept the secret until after 5 June 1793 when she wrote quite confidently to Jane, 'I have obtained my

Uncle's consent to my visiting my dear Friends at Halifax' and renewed her invitation to 'visit me and find me united to my dear William'. Filled with plans, she had only time 'to give you the bare information of my Happiness'.[16] Eleven days later she wrote again, with a total and dramatic change of mood. 'I cannot foresee the Day of my Felicity, the Day in which I am once more to find a Home under the same Roof with my Brother; all is still obscure and dark and there is much Ground to fear that my Scheme may prove a Shadow, a mere vision of Happiness.' William's last visit in the winter of 1790 seemed like a paradise lost; so did their walks, which she now coloured with emotion, 'when the keenest North Wind has been whistling amongst the trees over our Heads. . . . Ah! Jane! I never thought of the cold when he was with me.'[17] The 'little Parsonage', of which she had dreamt for two and a half years, had collapsed like a house of cards. Uncle William was not the comfortable eighteenth-century type of clergyman to take a bastardy in his stride; moreover he was harassed, since his wife gave birth to their fourth child on 14 June and the whole household was upset. Nor would that prospective patron, Mr Robinson, MP, have been pleased. Not only was there no question of a curacy for William: his uncle refused even to invite him to visit Forncett; 'he is no favourite with him alas! alas!' lamented Dorothy. For a short time even her hopeful spirit seemed to fail.

Then this young woman, who had been moved like a pawn by relatives and guardians all her life, began to show signs of unexpected character. She had always been open, transparent in her feelings, yet faced with the loss of William and home, she learnt to plot in secret. Outwardly she remained docile. Yet inwardly, with growing determination, she prepared for a new life. Her plan was sketched, with dramatic underlining, in a letter to Jane of 10 and 12 July 1793. '*None of this is to be read aloud, so be upon your guard*!'[18]

Since she had no hope of seeing her brother at Forncett, she formed in spite of 'a conflict within my breast a scheme which we have in agitation of bringing about a meeting at Halifax'. William had a standing invitation from the hospitable Rawsons. Dorothy had her uncle's permission to visit them but if he knew William would be there, he might refuse to let her go. Dorothy did not explain to Jane the reason for her uncle's objection to

her brother; 'the subject is an unpleasant one for a letter . . .
though I must confess that he has been somewhat to blame'.
She herself would not say a word, even to Aunt Rawson, of
'William's intention to visit Halifax while I am there', and
Jane must keep the secret—'Do not mention, I entreat you
again, my plan of meeting William'[19]—or Uncle William would
know 'that the scheme was a *premeditated* one'. Dorothy even
allowed herself to criticize this dependable relative, a wholly
unheard-of idea. 'He is one of the best of men but extremely
indolent (this pray never mention) and does not blend much
instruction with his conversation or enter much into my
studies.' This was less than just to William Cookson, who had
taken a BD and was now reading for a doctorate. Dorothy had
loved him, but liberty was in the air and she was entering upon
her own Revolution. Characteristically, it would be domestic
in scale, feminine, scrupulous to hurt no one. Yet the decision
to champion William was as decisive in her life as the Fall of
the Bastille.

For the moment nothing remained but to wait upon the plans
of Mr Griffith, Aunt Rawson's cousin from Newcastle, who was
to visit London on business and escort Dorothy to the North. She
waited, in fact, for six months while 'my secret remains a secret',
going through her household duties in an atmosphere heavy
with disapproval of William. Her aunt Cookson was particularly kind to her, giving her a present of five guineas towards
the holiday, perhaps in return for Dorothy's devoted care at the
birth of the children. Even this turned out badly, for Dorothy
lost her purse changing a guinea at Long Stratton and was forced
to write in August 1793 to Richard, 'it hurts me much to apply
to you in this distress, but what can I do? I have not a farthing
in the world.' Richard promptly sent her a ten-pound note,
debiting it to her account.[20]

William's future was a source of nagging anxiety. She saw that
the collapse of the curacy plan left him without prospects. 'I am
very anxious about him just now as he has not yet got any settled Employment,' she had written in June 1793, as soon as her
uncle's offer was withdrawn. The post of 'Tutor to some young
Gentleman, an office for which even Friends less partial than I
am, allow him to be particularly well qualified',[21] was filled by
someone else, much to her disappointment.[22] Instead, after a

South of England holiday with his school-friend, William Calvert, William set off on a solitary walk across Salisbury Plain, Bristol, and the Wye Valley into North Wales, composing the poem which later became *Guilt and Sorrow*. In calmer days he remembered himself on this walk

> more like a man
> Flying from something that he dreads than one
> Who sought the thing he loved.

It was probably at this time that he seemed to his friends to be threatened with a breakdown, which Dorothy was always later reluctant to mention. Perhaps, mercifully, this anguish of mind was not yet fully known to his sister, who imagined him spending the summer of 1793 with his college friend Jones in 'the most delicious of all Vales, the Vale of Clwyd' and passing his time 'as happily as he could desire; exactly according to his Taste, except alas! (ah here I sigh) that he is separated from those he loves'.[23] In autumn and winter he moved restlessly among friends and relatives in Cumberland. 'I have been doing nothing and still continue to be doing nothing. What is to become of me I know not,' he wrote.[24] Dorothy was forced to wait at Forncett.

This time of waiting was crucial for Dorothy Wordsworth's character. In the crowded, busy parsonage she was lonely, feeling 'a painful idea that one's existence is of very little use, which I really have always been obliged to feel'.[25] She must show nothing of her distress. She made resolute preparations, 'rising betimes in the morning and labouring as if to earn my Bread'. She grew pallid, 'wishy-washy' as she called it, and weary, unable to walk up the stairs to her garret sleeping-room without throwing herself breathless on the bed. She forthrightly put down her lassitude to 'the worms', took medicine, and forced herself to recover. She begged Aunt Rawson for news of Mr Griffith's plans: 'I am so very anxious that I know not how I shall support any further delay.' To Jane she poured out in the language of sensibility all the passionate feelings which were pent up in everyday life: how she would 'palpitate with rapture when I once more throw myself into your arms'.[26] 'I am sad, very, very often,' she confessed. The waiting had been so long. 'You can have no idea of my impatience to see this

dear Brother. It is nearly three years since we parted.' The
summer, autumn, and winter of 1793 seemed to her to crawl
by. 'Oh count, count the Days . . .,' she wrote; 'how slowly
does each day move! . . . long, long Months. I measure them
with a Lover's scale.' Yet never in a sentence, a line, or a
word did she give up her plan of a future with William.

Early in February 1794, with or without Mr Griffith, Dorothy
set off on a complicated and expensive journey. She took a two-
wheeled whiskey to Long Stratton, where the twice-daily Norwich to London coach stopped,[27] and was carried, at a cost of
one pound and ten shillings, probably to the Swan with Two
Necks, Lad Lane in the City.[28] A Halifax coach left daily from
the Bull's Mouth, Aldersgate, and the Saracen's Head, Snows
Hill. The journey was twice as long and the fare roughly double,
plus the usual charges for luggage, lodging, shilling breakfasts,
and well-earned rum and water for the coachman and guard. It
is hard to see how Dorothy could have paid this from her
minute income.

At Halifax it was a pleasure to see 'dear Aunt . . . in a situation so much more suited to her inclination and merits' than
the crowded shop in the town. Mill House with its cherished
garden was three miles out of Halifax on the banks of the river
Ryburn, almost in the country. William Rawson, Unitarian, a
founder of the Piece Hall cloth exchange and member of one of
the town's great textile dynasties, was kind and generous to his
new niece. Dorothy's happiness was complete by 7 February
1794, when she and William, after more than three years' separation, were once more 'under the same roof'.[29] They stayed
for six weeks, during which she learnt the troubled state of her
brother's mind and the depth of his need for her. To the kindly
and liberal Rawsons he appeared an eccentric young man.
Dorothy had been anxious about the impression he would create
at first, 'rather plain than otherwise', and stiff in conversation
with strangers.[30] His Republican principles were against him
while the Terror raged in France. In May 1794 the Habeas
Corpus Act was suspended and members of the radical London
Corresponding Society were arrested for 'treasonable practices'. William openly sympathized with them. Richard warned
him to be 'cautious in expressing or writing your political opinions. By the suspension of the Habeas Corpus Acts, the Min-

isters have great power.'[31] He also charged William to burn his
letter of warning because Lord Lowther 'has so many spies in
every part of the country'. This was realistic warning, and
Dorothy did not flinch from the dangers of her brother's position. 'I think I can answer for William's caution' about expressing his political views. He 'seems well aware of the
dangers of a contrary conduct'.[32] Indeed, though self-styled as
'of that odious class of men called democrats',[33] William's
opinions were, in fact, comparatively innocuous. He recoiled
'from the bare idea of a revolution', he was 'a determined
enemy to every species of violence', and he deplored 'the miserable situation of the French', still in the grip of the Terror.[34] All
the same, for Dorothy apparently to accept William's beliefs,
especially his violent anti-clerical bias, was a strange, swift
metamorphosis of a young woman, who, less than a year before,
had planned life in a well-ordered parsonage.

Their new life was marked by a journey together. It concentrated the pattern of their future. They took coach to Kendal,
and set off to walk the thirty-three miles through the Lake Country to Keswick, 'Two glad Foot-travellers through sun and
showers'. The early April weather of 1794 in the North-West
was dark and stormy,[35] yet Dorothy, tramping through the mud
to Windermere, 'with my brother at my side' felt it 'the most
delightful country that ever was seen'. The details of the walk
were fresh in her mind years later. 'I am always glad to see
Stavely it is a place I dearly love to think of—the first mountain
village that I came to with Wm, when we first began our pilgrimage together. Here we drank a Bason of milk at a public house
and here I washed my feet in the Brook, and put on a pair of
silk stockings by Wm's advice.'[36] They came to Grasmere at
sunset, when the sky cleared. 'There was a rich yellow light on
the waters and the islands were reflected there.'[37] Next morning they pushed on the remaining fifteen miles to a stone farmhouse in a lonely lane, Windy Brow on Latrigg, high above
Keswick. This farm belonged to William's school-friend William Calvert, who lodged there with his younger brother
Raisley; the young men moved into a nearby cottage to make
room for the Wordsworths.

Dorothy was delighted with the tenant farmer and his wife,
simple, honest, sensible, '*happier* than anybody I know'. She

calculated 'for how very small a sum we could live . . . our
breakfast and supper are of milk and our dinner chiefly of
potatoes and we drink no tea'. In mid-April the weather
became sunny and warm . The view up Borrowdale to the
south was magnificent: 'we command a view of the whole
vale of Keswick (the vale of Elysium, as Mr. Grey calls it)'.[38]
Dorothy, writing this to Jane, has not yet found her individual
touch for natural description. She not only quotes Thomas
Gray's *Journal in the Lakes*, but goes on much in Gray's style.
Her enjoyment, though, was spontaneous. She had intended to
stay only a few days, but the country was so beautiful that the
days turned to weeks,[39] as they explored the rocks and streams
or read Italian together.

   This gypsy interlude did not pass unnoticed by strait-laced
relatives at Penrith. Dorothy received a severe letter from her
uncle Christopher's wife,* complaining of her 'rambling about
the country on foot', and implying that she was 'in an unpro-
tected situation'. On 21 April 1794 Dorothy sent an answer,
carefully copied, quite unlike her usual impulsive scrawl to
friends. 'I am much obliged to you for the frankness with which
you have expressed your sentiments upon my conduct,' she
wrote, 'and am at the same time extremely sorry that you should
think it so severely to be condemned.' Point by point, she
defended herself. She was living simply and cheaply: 'I drink
no tea', then a luxury on which duty was paid. Good tea was
expensive enough to sell by the ounce, and used tea-leaves
were sold at the back-doors of houses. As to chaperonage, 'I
affirm that I consider the character and virtues of my brother as
a sufficient protection'. Walking 'not only procured me infinit-
ely more pleasure than I should have received from sitting in a
post-chaise—but was also the means of saving me at least thirty
shillings', a shrewd blow. She was improving her education
with William. Moreover, 'I am now twenty-two years of age
and such have been the circumstances of my life that I may be
said to have enjoyed his company only for a *very few* months'.
She could not lose it, 'without unspeakable pain'.[40] In conclusion, with civil thanks, she accepted an invitation to visit her

---

   * Christopher Cookson had changed his name to Crackanthorpe in 1792, when he
inherited the Crackanthorpe Estate through his mother.

aunt and uncle. This even-tempered but implacable reply marks the appearance of Dorothy Wordsworth, no longer a ser- viceable poor relation, but an independent woman.

She passed the next fifteen months in a series of visits to rela- tions, some of whom had not seen her since childhood, a tribute to the family tie in this large North-Country clan. The first visit, which included William, was in May 1794 to their guar- dian Uncle Richard Wordsworth, at his son's house, Branth- waite near Whitehaven. The road lay through Cockermouth, where they stopped to look at the house of their birth, which Dorothy had last seen when she was six years old. Now it had stood empty for many years; 'all was in ruin, the terrace-walk buried and choked up with the old privet hedge . . . the same hedge where the sparrows were used to build their nests'.[41] Nothing could have reminded her more sharply of their essential homelessness and her longing for her own brothers. She wrote urgently, begging Richard to press the Lonsdale suit and the settlement of their father's estate, 'daily more distressing and intricate. . . . These things make me very uneasy.'[42] Richard replied rather stiffly to a further letter from William, 'allow me to assure you that I have always had my Sisters, yours and my younger Brothers interest at Heart, although I have not been fond of making professions which it could not be my intention to carry into effect'; he would do what he could, but the law must take its course.

At Branthwaite, Dorothy's aunt, 'a sweet woman', welcomed her kindly and gave her free use of her own time, a luxury after so many domestic duties. But her uncle was ill with jaun- dice and dropsy and 'the physicians here seem to have no hopes'. In mid-June he died, appearing in the 'obituary of con- siderable persons' in the *Gentleman's Magazine* of July 1794. Dorothy and William left the mourning household. Lacking home or money, their holiday together was ended and they were to be parted again for more than a year. William returned to the Calverts at Keswick, while Dorothy rode her uncle's horse to another cousin, Mrs Barker at Rampside, on the coast near Furness Abbey, and after 'about a fortnight' went to stay with the Speddings at Armathwaite. These were the family of a Hawkshead Grammar School friend of William, kind to her as they had always been to him. Coleridge later found them 'nothing remarkable in minds or bodies . . . chatty, sensible women, republican in opinion and just like other Ladies of their rank in practise'.[43] To Dorothy they seemed 'in every respect charming women . . . have read much and are amiable and engaging in their manners. They live in the most beautiful place that ever was beheld.'[44]

In August 1794 Dorothy moved on, as family duty demanded, for her promised visit to Aunt and Uncle Crackanthorpe at Newbiggin Hall, about six miles from Appleby. Here a surprise awaited her. Uncle Christopher, the imagined ogre-figure of Penrith days, was most affectionate and strangely moved: 'I never saw a man so agitated in my life as he was at our meeting'. Did Dorothy at twenty-two perhaps resemble her mother, his dead sister? 'I could not resist and in any case I am very glad that we had a meeting.' He gave her ten guineas, a very substantial sum; not only was this kind, but 'the manner of giving it made it doubly so'.[45] Dorothy grew convinced that he would always have loved her, but for that 'proud and selfish woman' his wife, and it appears that she stayed with them for some months. Christmas and the New Year she spent with the Griffith family at Newcastle.[46] It was an iron-cold winter. Across the North Sea, the river Texel froze, and French hussars rode over the ice to capture the fleet of England's ally, the Dutch. In Newcastle, more prosaically, the chamber-pots froze above stairs. Yet the cold did not affect her enjoyment and she found 'our good friends . . . very cheerful pleasant companions'. She joined in 'all the Newcastle adventures' and did not leave them until early April 1795.

Dorothy had intended to go, sharing a chaise with cousin Elizabeth Threlkeld, directly back to the Rawsons at Halifax. Then she remembered her friends, Mary and Peggy Hutchinson, with whom she had 'compared grievances and lamented the misfortune of losing our parents at an early age', in the unhappy days of her girlhood at Penrith. They now kept house for their brother Tom on a farm which he had inherited from his grandfather at Sockburn, near Darlington. Impulsively, two days before leaving Newcastle, Dorothy wrote proposing a meeting, and inevitably they persuaded her to stay. The farm was 'washed nearly round by the Tees (a noble river). . . . We spend our time very pleasantly in walking, reading, working and playing at ball in the meadow in which the house stands, which is scattered over with sheep and "green as an emerald".' It was a joy to see them so happy, 'exactly as our imaginations and wishes used to represent'. If only she had such a home—'but these are airy dreams'.[47] She left the Hutchinsons, feeling she might never see them again: a remarkable miscalculation as it turned out.

At Northallerton she took the cheaper 'heavy coach' back to Halifax and Mill House. The Rawsons made her as welcome as ever, and did everything they could to make her time pass agreeably. The cousins were all delighted to see her; there were dinners at the Threlkelds and tea-drinking with Patty, now promoted to be Miss Martha Ferguson, manageress of the haberdashery shop. They discussed the wartime guinea-a-head tax on hair powder and Dorothy declared her 'disapprobation of the present destructive war',[48] against the French republic. Cousin Edward Ferguson also opposed the war and cropped his hair 'à la guillotine' rather than pay the tax. Such radical opinions seemed permissible at Mill House; sympathetic Uncle Rawson bought her two elegantly bound volumes, Akenside's *Pleasures of Imagination* and Rogers's *Pleasures of Memory*, both edited by Mrs Barbauld, Unitarian verse-writer and pamphleteer in the rights of man. The high point of summer came on 5 August 1795, when Dorothy was present at Jane's wedding to John Marshall of Leeds, congratulating her 'with the most heartfelt pleasure'. Mill House with its large flower garden and green lawn was so delightful that she grieved at the thought of leaving it; as she later wrote, with touching simplicity, 'Mr. and Mrs. R. have been so *very*, very kind to me'.[49]

In late August 1795, Brother Kit came to stay for a week, all the time he could spare from hard study at Cambridge, since he graduated Tenth Wrangler the following winter. His disposition proved 'delightful . . . to tell you that he is very like me may make you smile, as it is a little like vanity; it is, however, allowed by everyone, and I myself think I never saw a stronger likeness'. This raises the question of Dorothy Wordsworth's looks as a young woman in her twenties. A silhouette exists which must belong to early years, before the drastic effects of illness, work, and loss of teeth upon her profile. Christopher's high forehead, aquiline nose, and firm mouth were strong in character.[50]

The same features appear in a portrait of Aunt Rawson, set off by immaculately frilled lace collar and cap; her expression is composed and enquiring.[51] All three members of the family shared the same deep-set eyes, lively and appealing; Dorothy's, were a flashing, glancing dark grey. In build she was small and slight; in manner ardent, in movement swift and spontaneous.

Happy as she frequently was, among friends and relations, the main and immediate aim of her life was in suspense, up to 1795. It was a year since she had contrived to escape from the kindly Cooksons to meet William, and she had still not set up a home with him. The object she had so often outlined to Jane was still not achieved, nor did it look likely to be in the foreseeable future. It was not just a matter of persuading sceptical relatives about 'the character and virtues of my Brother'. Even after all her efforts, his position in the world offered no hopes of a family home.

5. 'The First Home'

Dorothy's time of suspense was ended by an intervention of fortune, vital to her and William. William Calvert's younger brother, Raisley, no poet himself, nevertheless became convinced of Wordsworth's genius. Raisley was due to come into a trust fund of his dead father later in 1794. He took Wordsworth to live with him in June 1794, and, apart from a month's break, he stayed with Raisley for the rest of that year. It became clear the young man was in an advanced state of consumption, and he determined to subsidize the poet's genius by leaving him a legacy of £600 (later £900), so that he could devote himself to poetry. A codicil to Raisley's will gave William powers to invest 'for the use and benefit of his Sister', so that Dorothy could housekeep for her brother. Raisley died on 9 or 10 January 1795. His recent will, which he had signed on 23 October 1794, was proved, and its provisions set in motion. From 1795, the annual income of brother and sister provided a bare basis for livelihood, independent of the charity of relatives, which had irked so much—'Above all', admitted Dorothy in September 1795, able to speak freely at last, 'it is painful when one is living upon the bounty of one's friends'.[1]

Both Calvert brothers were young men of generous and unconventional mind. William Calvert, though holding a commission as a militia officer, who might be used in punitive actions against just such 'democrats' as Wordsworth, never lost his admiring friendship with the poet. Even more remarkably, the twenty-one-year-old Raisley spent his dying months safeguarding the legacy for Dorothy and her brother. Realizing that Wordsworth owed over £400 in educational expenses to his late uncle, Richard Wordsworth, and that his uncle's heirs might claim repayment from Calvert's legacy, Raisley insisted that a legal bond be drawn to prevent this happening.[2] It is hard to think of such selfless devotion in the face of imminent death.

Even the legacy did not at once solve Dorothy's and William's problems. The income raised would be minimal, while, it must be realized, William was left paying interest on what he still owed his uncle's estate.[3] It was a war period of rising prices, and throughout the following few years they had to practise the strictest economy. Dorothy had to make do with adapting cast-off clothes for both of them, and William had at all costs to make any garden they might possess support them with vegetables and fruit—'To drink wild water and to pluck green herbs'. Moreover, the Calvert legacy was only paid over slowly during the next few years. It was not until 13 August 1798 that brother Richard had the last instalment in his hands to invest safely in 3 per cent Consols.[4] Nor was this need of steady income helped by William's various unorthodox means for obtaining higher rates of interest.[5] In any case, Dorothy and William, however secure in their minute income, still lacked the first essential, a home. Once again, unexpected fortune at last appeared to favour them, just when Aunt Cookson, in summer 1795, was expecting her fifth child in less than six years, and wanted Dorothy back at Forncett to help in the overcrowded nursery, where 'the mind is unfitted, perhaps, for any new exertions and continues always in a state of dependence'.[6]

The new opportunity of independence came through another chance friendship of William's. While nursing Raisley Calvert in Cumberland, William had become temporarily convinced that 'cataracts and mountains will not do for constant companions'. He carried this conviction to such a length that he wished Dorothy to join him in London, where they would try to augment their new incomes by writing and translation. The scheme, forthrightly opposed by 'Aunt' Rawson, in a letter to Sam Ferguson, as 'a very bad wild scheme',[7] luckily came to nothing. Yet William's six-month stay on his own in London, February to August 1795, bore its desired result for Dorothy and himself, a home of their own. In London, he seems to have lodged not with his brother Richard, but with a young and widowed law student, Basil Montagu, who had been an exact contemporary at Cambridge. Montagu took pupils, among whom were the two sons of a rich West India merchant of Bristol, John Pinney. One son, John Frederick, offered William a West-Country retreat. Pinney senior, fearing his West Indian interests threatened by the American war, had refurbished, as 'a lee-port in a storm', a family property on the Dorset-Devon border, renaming it Racedown Lodge. John

Pinney suggested the place, empty apart from the caretaker, a poor relation called Gill, as the Wordsworths' home. The widowed Montagu had a two-year-old son, whom William could bring up in the country, nominally coaching him, though naturally the effective care would fall on Dorothy, in return for money to supplement their joint income.

Dorothy was enthusiastic about the scheme, though she faced her responsibilities honestly. 'It will be a very great charge for me,' she wrote to the newlymarried Jane Marshall, 'but it is of a nature well suited to my inclinations. You know I am active, not averse to household employments and fond of children.' When she thought of her duties she was 'anxious and sometimes fearful', but determined to do her best. She consulted Uncle William, who approved. Kit felt happy that he and John would have 'a place to draw to'. The Rawsons clearly felt that Dorset was safer than Grub Street for their niece. 'My Aunt says she has no doubt of me and Mr. Rawson is of the same opinion. He thinks I am quite equal to the charge.' Above all, it would be a 'permanent establishment', after the months of drifting; 'it will greatly contribute to my happiness and place me in such a situation that I shall be *doing something*'.[8]

William went to Racedown via Bristol, waiting in that city for Dorothy to join him from the north. The coach connections were good; Dorothy travelled from Halifax to Birmingham and on to Worcester, where she took coach to Bristol and met William on 22 September.[9] 'I like her manner and appearance extremely, she is very animated and unaffected,' wrote the young Pinneys' elder sister.[10] On 26 September, Joseph Gill, caretaker at Racedown Lodge, recorded in his diary, 'At midnight arrived Mr. and Miss Wordsworth'.

Racedown Lodge, a square Georgian mansion of dignified red brick, with soaring chimneys, stood on a hillside outside the hamlet of Birdsmoorgate, between Lyme and Crewkerne, the local post town. It was a gentlemanly country seat for an impoverished poet, its garden walks decorated with statues, or 'images', as Gill the caretaker called them; but the Pinney brothers offered it rent-free, and it became the first house where Dorothy was her own mistress. Large as it was, she ran it single-handed for a month, until 'one of the nicest girls I ever saw', Peggy Marsh, came as general servant and devoted

friend to the Wordsworths. She and a washerwoman once
a month at ninepence a day constituted the whole staff. Her
wages are not mentioned, but £6-£8 a year 'with diet' was
usual in the 1790s,[11] and both parties seemed content. The vast
cavernous kitchens had no terrors for Dorothy, who soon had
'all my domestic concerns so arranged that everything goes on
with the utmost regularity'.[12] She was deeply happy. 'I think
Racedown is the place dearest to my recollections upon the
whole surface of the island,' she wrote some years later. 'It was
the first home I had', and she recalled 'the lovely meadows
above the tops of the combes' which she could see from every
window of the house.[13] On either side of the elegant staircase
were two front parlours, and two smaller rooms looked over the
kitchen garden at the back. Here William, wrote Dorothy,
'handles the spade with great dexterity'.[14] A gardener came to
grind the orchard apples for cider, William hewed wood, and
Gill, in the soft spring weather of 1796, 'planted one bed cab-
bage plants, one bed pease', essential to the Wordsworths, who
needed to live as cheaply as possible on their own produce.

The high cost of coal, which came by sea to Lyme and had to
be fetched by a neighbouring farmer's cart, was an anxiety to
Dorothy. 'You would be surprised to see what a small cart full
we get for three or four and twenty shillings';[15] their poor
neighbours were reduced to stealing the garden fences for fuel.
In cold weather 'our little breakfast room' became their parlour.
From old Mr Pinney's inventory, the 'very neat furniture' included a pianoforte, a mahogany sofa and armchairs, a Bath
stove, a painted floor-cloth, a picture of Venice, and above all,
to their delight, two mahogany glass-fronted bookcases filled
with books, Italian, English poetry, science, and Puritan theol-
ogy. 'We are now surrounded by winter prospects outdoors',
wrote Dorothy to Jane at the end of November 1795, 'and
within have only winter occupations, books, solitude and the
fire side, yet I may safely say we are never dull.'[16] In fact, the
'winter prospects' in the South of England hardly seemed so to
a Northerner like Dorothy, who found it the mildest winter she
ever remembered.[17]

She studied Italian 'very hard' and read widely among the
Pinneys' books: travel, Henry Brooke's *The Fool of Quality*,
'which amuses me exceedingly', Madame Roland's *Memoirs*,

and *Tristram Shandy*, like her girlhood reading of *Clarissa* a Hanoverian taste, suppressed during the pious rectory years. Her Italian extended to reading Ariosto with William. The little boy Basil, not three years old when he came to them, 'affords us perpetual entertainment', though Dorothy would not allow herself to treat him as a plaything, for fear of harming his temper. He had lost his mother at birth and had lived either a confined existence in his father's chambers at Lincoln's Inn, or farmed out erratically among his father's friends. He arrived 'extremely petted from indulgence and weakness and perpetually disposed to cry'. Jane Marshall, herself now the mother of a baby boy, wrote to ask after Dorothy's 'system'. 'It is a very simple one,' replied Dorothy. 'We teach him nothing but what he learns from the evidence of his senses. He has an insatiable curiosity which we are always careful to satisfy to the best of our ability. It is directed to everything he sees, the sky, the fields, trees, shrubs, corn, the making of tools, carts, etc., etc., etc. He knows his letters, but we have not attempted any further step in the path of *book learning*. Our grand study has been to make him *happy*.'[18] To Aunt Rawson Dorothy made the same point. 'Till a child is four years old he needs no other companions than the flowers, the grass, the cattle, the sheep that scamper away from him when he makes a vain unexpecting chase after them, the pebbles upon the road, etc., etc.'[19] These were the principles of Rousseau's *Emile*, current in the radical circle William had met at the bookshop in St. Paul's Churchyard: as in *Emile* Book Two, an education in the country 'according to nature' and 'in no hurry to teach'. Basil was required to learn without punishment, 'by experience alone'; if he would not get up when called, no one dressed him, and he was 'obliged to go to bed again where he lay till 4 o'clock'. Like the conveniently imaginary Emile, the living Basil seemed to profit by this treatment. After six months he changed 'from a shivering half-starved plant to a lusty, blooming, fearless boy', who played outdoors in all weathers, heedless of wind or rain. Dorothy watched his development with endless interest; 'he is my perpetual pleasure'.[20] All her life, she enjoyed the company of children.

Dorothy and William took full advantage of the mild Southern climate. Every morning they walked for about two hours,

over the sandy hillsides, quick-drying after rain. They had a
view of the sea, about two hundred yards from their door,[21]
and a short distance from the garden gates a track climbed to
the earthworks on Pilsdon Pen, the highest point in Dorset.
Here, in November as she described it, the countryside has
hardly changed since Dorothy took it to her heart. The great
slopes of Marshwood Vale open at the walker's feet, sheep
browse near their dew pond, the yaffle shouts from the deep
wooded coombe, and distant hills open to the sea. 'We have
hills which seen from a distance almost take the character of
mountains,' she wrote to Jane on 30 November 1795, 'some
cultivated nearly to their summits, others in their wild state
covered with furze and broom. These delight me the most as
they remind me of our native wilds.'[22] Yet, years later in these
'wilds', she still recalled walks on Pilsdon, Lewesdon, Blackdown, and the earthwork called Lambert's Castle.[23]

One quiet country day followed another, apparently unevent-
fully, yet filled with inner life. A letter came from France, the
only one of six which Annette had written to arrive; Caroline,
William's daughter, was now nearly three years old. As a
reminder of the war which parted them, William saw, looking
out to sea, the English fleet for the West Indies, fated to meet a
hurricane, which strewed the nearby Dorset coast with corpses.
Inland, they walked the seven miles to Crewkerne to collect
their letters from the post office, Dorothy adapting her short
quick steps to William's stride; once they went further to admire the fine view from Lord Powlett's house at Hinton St.
George, and 'were amply repaid for their trouble'. Their nearest neighbours, another branch of the Pinney family from
Blackdown, called on them. They were 'a very good kind of
people', wrote Dorothy, and willing to be helpful, yet 'they
have not much conversation'. Their society did not promise
entertainment.

New Year 1796, was enlivened by a week's visit from the
Pinney brothers of Bristol, who returned for most of February.
'We all enjoyed ourselves very much,' reported Dorothy to
Jane. The weather continued so mild that the North-Country
Wordsworths could hardly believe it was winter, until one late
snowfall in the first week of March. The young men were out
all the morning, hunting, coursing, clearing wood which

3. 'upon the Terrace . . . I fancied myself treading upon Fairy Ground': Windsor Castle

4. 'woods as wild as fancy ever painted': Alfoxton Park and the Quantock Hills

'produces warmth both within and without doors', or walking
in which Dorothy joined. In the evenings they seemed 'to relish
the pleasures of our fireside'. Riches had not spoiled them;
they were well-educated, well-informed, and 'very agreeable in
conversation'. By agreement they paid board during their visit
and Dorothy dutifully invited the other neighbouring Pinneys
to dine with them, although it was not really her style. 'This
was what we call a *grand* rout and very dull it was except for the
entertainment in talking about it before and after.'[24] A crisis
was averted when old Mr Pinney in Bristol discovered that the
Wordsworths were paying no rent and threatened to charge
them, but 'at our earnest desire' his sons persuaded him to
relent.[25] They were not, as it happened, in the least interested
in poetry; simply, like Raisley Calvert, they had fallen under
the spell of Wordsworth and his sister.

When they left, Dorothy settled down, on 7 March 1796 with
unexpected snow on the ground outside, to answer four letters,
including those of her aunt Rawson and Jane. 'We seem quite
quiet'; yet they were never bored, for they were absorbed, endlessly interested, in each other's society. Racedown was the
scene of events which were supremely important in the lives of
Dorothy and her brother. In the latter's words,

> She, in the midst of all, preserv'd me still
> A Poet, made me seek beneath that name
> My office upon earth, and nowhere else.

There have been many explanations by biographers of these
words, written nearly ten years later, in *The Prelude*, but referring to this time. Remorse for Annette, wavering faith in
human nature, doubt about Wordsworth's own calling as a
poet, have been suggested as the handicaps Dorothy had to
clear away from the troubled state of her brother's 'moral
being'.[26] Yet perhaps the most vital gift of hers was that, as he
afterwards acknowledged, she taught him to see life through
her eyes, and specially to see Nature as it was, and not through
a curtain of literary allusion. Her own words, as she continued
this letter to Jane, illustrate how this could be. Whereas William, in another letter, described his walk past Lambert's
Castle towards the sea at Lyme entirely in terms of literary
reference, by quoting from his favourite but well-worn poem,

James Beattie's *The Minstrel*, Dorothy uses her own simple, accurate, yet imaginative phrases. Continuing to describe the wide view seaward from the Iron Age enclosure at the summit of Pilsdon Pen, with its barrows and double ramparts, half-masked by stunted trees that 'suffered from the sea-blasts', she looked out and observed 'a very extensive view terminated by the sea seen through different openings of the unequal hills'.

The end of that sentence shows, perhaps for the first time, Dorothy's own alert poetic way with words. To begin with, it is, quite simply, a blank verse line,

> Through different openings of the unequal hills.

Yet, although a kind of poetry, it is as factually descriptive as prose. The superb view due south, over Lyme Bay, is seen through coastal hills, lower than the commanding Pen itself, but each one of a different, 'unequal' character. The sun-burnished water glistens like bronze metal through gaps of every shape. Some hills, running toward steep cliffs, such as flat-topped Golden Cap, are sharp and rectangular; many others are conical, miniature isosceles triangles, often with pheasant-shaped crests of small pine plantations surmounting their pointed sandstone summits. 'Through different openings of the unequal hills' is not only a plain and straightforward description, it is, as poetry, a line remarkably like the type of blank verse William was to write three years later when he began *The Prelude*. This is not to say that he learned to write a new kind of poetry from Dorothy, a poetry which we now recognize as 'Wordsworthian'; but she pointed him the way in her own prose, written, like her later journals, with no premeditated art, but from a deep loving identification and sympathy with the place or thing described.

The same attitude informs her remarks about people. Both she and William were startled by the state of the Dorset peasantry, 'miserably poor', 'wretchedly poor'. Whereas William used this as an occasion for a short burst of Godwinian philosophy, tracing the misery to ignorance and connecting it with evil—'ignorant and overwhelmed with every vice that usually attends ignorance in that class'—Dorothy paints a more human and understanding picture. In modest terms, she concentrates not on the evils of their nature, but the handicaps of

their environment: 'their cottages are shapeless structures (I
may almost say) of wood and clay—indeed, they are not at all
beyond what might be expected in savage life'.[27] This was no
more than the truth. Farm labourers lived on 'an unvarying
meal of dry bread and cheese from week's end to week's end',
while over the Devon border, in an enclosed village without
commons, not one working man could maintain his family on
his shilling-a-day's wages.[28]

Moreover, to share an isolated life with one so sensitive and
responsive brought a new tenderness to William's own activity.
He begins to anticipate what she would like, what would interest her, the sort of experience that would bring her pleasure.
The iron casing that had cramped his life with anxiety begins to
fall off. Small, tender, domestic incidents, in which he forgets
himself, and thinks only of her, take their part in the day-to-
day conduct of his life. One of these is commemorated in a
poem written half-a-dozen years later, but which describes an
actual happening at Racedown round about mid-July of the
year 1796.* William had the habit of riding a horse into Lyme
and back. Indeed, a local legend, still current earlier in the present century, has it that he once rode in and absent-mindedly
walked home the eight and a half miles through Uplyme to
Birdsmoorgate, leaving the horse still tethered back in Lyme.[29]
In the poem, alive with the delight of anticipating how he can
please Dorothy, he describes himself as taking special pains
over the smallest incident on one of his return journeys from
Lyme, late 'one stormy night' in summer.

> Among all lovely things my Love had been;
> Had noted well the stars, all flowers that grew
> About her home; but she had never seen
> A Glow-worm, never one, and this I knew.
>
> While riding near her home one stormy night
> A single Glow-worm did I chance to espy:
> I gave a fervent welcome to the sight,
> And from my Horse I leapt; great joy had I.
>
> Upon a leaf the Glow-worm did I lay,
> To bear it with me through the stormy night;

* William wrote in 1802 that it occurred 'about six years ago'; the glow-worm
season makes it 1796, the first summer at Racedown.

And, as before, it shone without dismay;
Albeit putting forth a fainter light.

When to the Dwelling of my Love I came,
I went into the Orchard quietly;
And left the Glow-worm, blessing it by name,
Laid safely by itself, beneath a Tree.

The whole next day, I hoped, and hoped with fear;
At night the Glow-worm shone beneath the Tree:
I led my Emma to the spot, 'Look here!'
Oh! joy it was for her, and joy for me!

The incident was perhaps impressed on William's mind with special intensity, since he had just been absent from Dorothy for six weeks in London, returning to Racedown on 11 July. The poem's attention to detail makes one wonder whether the 'lovely things' of her solitary walks at that time had literally been 'noted well' by her in some now-lost journal; she certainly bought a 'Diary' for some purpose later that year.[30] Her 'joy' in the glow-worm suggests that Dorothy, as well as William, was changing in their shared life. From school days in the rational optimism of the Enlightenment, she had passed, at her uncle's rectory, to the intense personal service of the early Evangelicals. Now, love for her brother swept her into the visionary passion of early Romanticism, where every common sight was charged with a spiritual message. Her innate religious sense poured itself into the natural creation. By the time she began to keep her Journals, formal church-going had ceased to play a part in the Wordsworths' lives, and it seems likely that this change began at Racedown, since orthodox devotions are never mentioned in her Dorset letters.

Sad thoughts awoke at a 'melancholy letter' from Mary Hutchinson in spring 1796. Mary was nursing her sister Peggy on her deathbed from consumption. The happy visit at their brother's farm at Sockburn, the circling river, the pretty sheep and lambs, reappeared in a ghostly light; 'last year at this time we were all together and little supposed that any one of us was so near death'.[31] The following autumn, after Peggy's death, Mary was the first guest whom William and Dorothy chose to invite to Racedown; the Pinneys had come by right as owners. On about 28 November 1796 she arrived, escorted by her sailor-brother Henry, who left next day. Mary herself stayed for seven months. From the first, she fitted into the household. The Hutchinson family in many ways resembled the Wordsworths, without the streak of genius. Early orphaned and divided in childhood, they had made a family home for themselves by sheer determination; they were loyal, reserved, and in Mary's case, self-sacrificing to a fault. This is the only criticism which Dorothy ever made of her. Now the two young women settled to a shared daily life. 'My friend Mary Hutchinson is staying with me; she is one of the best girls in the world and we are as happy as human beings can be,' wrote Dorothy to Jane in March 1797.[32] They wrote letters at the same table, or sat together at their sewing: though no longer woven, all household linen was still sewn at home. Dorothy made frocks and shirts for small Basil, or mended William's distinctly shabby clothes. She asked brother Richard in London to pass on 'old clothes. They will be of great use at Racedown.'

Meanwhile Dorothy began on the heroic enterprise of Richard's shirts. This lawyer brother had thriftily bought a hundred yards of Irish linen and required his sister to sew him a dozen shirts, with stock and wristbands, 'a tolerable good length'. The cloth arrived at Racedown in March 1797, and Dorothy wrote asking for measurements, 'and whether you would choose *two* button holes upon the wrists'. Richard wrote rather peremptorily on 3 May, 'I will thank you to send me half a Doz. of the shirts as soon as they are finished. I am in great want', and was reassured on 7 May by William as Dorothy was ill, 'Your shirts are begun today and shall be despatched as fast as possible.' To anticipate the saga of the shirts, Mary Hutchinson delivered the first four in June. They proved too tight in the neck. 'It was a great pity you did not send me an accurate measure,' wrote Dorothy,[33] who had now made all but three, with creditable restraint. She altered all the neckbands and in January 1798 sent five more by another returning visitor, Basil Montagu. The shirts were not finished until the end of April 1798, when a last parcel was forwarded by goods wagon. For thirteen months they had formed an unquestioned ground bass to the music of poetry; for William, who had bought a half-share of the material, had to be provided with shirts too.

William was 'ardent in the composition of a tragedy', *The

Borderers,[34] which occupied him through the winter and spring
of 1796-7. Mary's company was welcome in womanly matters,
often so far removed from those of men. 'I am very happy in
her society,' Dorothy wrote to Richard. The two were as close
as they had been in girlhood, before Peggy Hutchinson's death.
They walked together through unfenced apple orchards, or
upon Pilsdon seeing the wooded slopes and tawny bracken of
the vale glistening in the wintry sun, or 'distant volumes' of
smoke rising from chimneys, 'when every cottage hidden from
the eye, pointed out its lurking place by an upright wreath of
white smoke'.[35] All three appeared contented in each other's
company. 'William is as cheerful as anybody can be,' wrote
Dorothy; 'he is the life of the whole house.' When in March
1797 Basil Montagu arrived unannounced before they were up
in the morning, the circle widened to admit him. On 18 March
he took William off for a two weeks' visit to Bristol. 'You cannot imagine how dull we feel,' confessed Dorothy to Jane on
19 March, 'but this is the first day; tomorrow we shall be
better.' She took it for granted, without possessiveness, that
Mary shared her feelings.

Dorothy had always assumed that some congenial friend of
her own would share her happiness with William. This had
been the picture she outlined to Jane Pollard, when she had
written to her nearly four years earlier in the summer of 1793.

. . . think of our moonlight walks attended by my own dear William,
think of our morning rambles when we shall . . . perhaps before
William rises . . . walk alone enjoying all the sweets of female friendship. Think of our mornings, we will work, William shall read to us.
Oh my dear friend how happy we shall be!

Now, with Jane married, and a mother, Dorothy transferred
these sentiments to her life at Racedown with Mary. Though
nothing in this matter can be certain, it seems likely that
Dorothy made Mary her confidante about William and Annette. Mary shared with Dorothy the copying of poems, written
by William at Racedown, which dealt, however indirectly,
with his feelings in the aftermath of the affair, such as parts of
The Ruined Cottage, and of a long ballad poem, *The Three Graves*.
There can be no final proof; but it seems highly likely that
among 'the sweets of female friendship' during the seven

months Dorothy shared with Mary full knowledge of Annette and William's child.

Legouis, writing on Annette, feels a gradual fading of William's passion, grief, and guilt towards her in the winter of 1796-7, a 'growing estrangement of which at first he may not have been quite conscious'.[36] While she was drawn into Royalist conspiracies and dangers, commitment to poetry steadily filled his life. Dorothy, who had calmed his first anguish, now fulfilled his need for feminine sympathy. She recalled the world of their childhood, from which he drew his deepest inspiration, their real world, which made the months in France seem like a feverish dream. To this world of childhood and home Mary Hutchinson belonged. Dorothy willingly re-entered it with them both, and, if she had other hopes or desires for herself, wrote nothing of them. So, over the months at Racedown, the bonds of a shared life for the three of them were gradually forged.

In spite of Dorothy's great happiness at Racedown, it would be mistaken to see the spring of 1797 as entirely carefree. The crisis over Annette, though fading in the necessary separation of the continuing war, cannot have been far from Dorothy's mind. There were further worries over health and money. The first mild winter of 1795 had been succeeded by the abnormally severe weather of 1796-7. The whole of December 1796 froze so hard that in many parts of the country the starving birds had stripped all the berries by Christmas. The icy cold lasted as late as May 1797, when William, returning from Bristol, wrote to Richard, 'Poor little Basil has been very ill. Dorothy has got a very bad cough, and I a terrible cold.' 'Poor Basil', wrote Dorothy, 'was very ill—I was afraid we should have lost him.' She herself had 'a violent cold' with high fever, swollen face, and raging toothache. By June she claimed to have recovered, but the illness had lasting consequences. She lost a stone and a quarter in weight, and three years later told Jane regretfully that she was thinner still.[37] In an age whose reigning beauty was the voluptuous Emma Hamilton, to weigh less than seven stone was a liability. Moreover, during the next five years her teeth were affected. 'My tooth broke today,' she wrote at the age of thirty, on 31 May 1802. 'They will soon be gone. Let that pass, I shall be beloved—I want no more.' Despite her

own lack of personal vanity or concern with outward appearance, by 1805 friends who had known her in Halifax days were
shocked at the change. In 1810 William reported hopefully,
'Her throat and neck are quite filled up; and if it were not for
her teeth she would really look quite young'; but the youthful
looks which had charmed Miss Pinney at Bristol had vanished
for ever. Worse than health was unexpected financial crisis.
Uncle Richard Wordsworth, the second trustee of the family
estate, had died in 1794. He had advanced over £400 for William's education, and though Raisley Calvert had secured his
own legacy to William from any claim, the son, Robinson
Wordsworth, was now pressing William for at least part repayment. Even with their tiny incomes boosted by interest from
the Calvert legacy, and even living rent-free, they could only
just manage at Racedown. Another optimistic scheme of Dorothy's, to take another child,[38] had failed to materialize. She
knew that Robinson's claims were just, if unpleasant, and
reflected badly upon William; in fact, 'that it was universally
reported in Cumberland that he had used his Uncle's children
very ill'. Dorothy had to admit, in natural honesty, 'their
claims are so just that it is absolutely necessary that something
must be done'.[39] Nothing *was* done, since William had no
available money, though he seemed, somewhat illogically, to
feel that his brother Richard could raise some for him.[40]

Their main preoccupation, however, was something quite
different. Mary Hutchinson, to Dorothy's regret, left Racedown for Crewkerne on 5 June 1797, bearing some of Dorothy's
shirts to leave for brother Richard on her way through London
back to the North; in return for this errand, Dorothy hoped,
Richard would show Mary some of the sights of the Town. On
the very next day, another visitor arrived at Racedown, making
a dramatic entrance on the road from Chard, whence he had
walked after an early breakfast at Taunton, where he had been
staying with the Unitarian minister. Seeing the Wordsworths
in their orchard, across the angle where the Chard and Crewkerne roads meet, he impulsively wasted no time going around
by the road, but, leaping a gate, plunged across the intervening
field, jumped the little stream, which formed the county boundary, and presented himself to them at Racedown. Such was
the characteristic advent of Samuel Taylor Coleridge.

6. 'Coleridge's Society'

'You had great loss', wrote Dorothy, almost at once, to Mary Hutchinson, 'in not seeing Coleridge. He is a wonderful man.' She went on:

His conversation teems with soul, mind, and spirit. Then he is so benevolent, so good tempered and cheerful, and, like William, interests himself so much about every little trifle. At first I thought him very plain, that is, for about three minutes: he is pale and thin, has a wide mouth, thick lips, and not very good teeth, longish loose-growing half-curling rough black hair. But if you hear him speak for five minutes you think no more of them. His eye is large and full, not dark but grey; such an eye as would receive from a heavy soul the dullest expression; but it speaks every emotion of his animated mind; it has more of the 'poet's eye in a fine frenzy rolling' than I ever witnessed. He had fine dark eyebrows and an overhanging forehead.[1]

Perhaps the most potent fact in this descriptive catalogue was that the newcomer 'like William, interests himself so much about every little trifle'. The thoughtful tenderness that had produced the incident of the glow-worm had another, new exponent. As for Coleridge's appearance, William was at one with Dorothy, and continued to be all through Coleridge's sadly changed life. After nearly another five years of sometimes-chequered partnership, Coleridge was for William, as he had first seemed to Dorothy,

> A noticeable Man with large grey eyes,
> And a pale face that seemed undoubtedly
> As if a blooming face it ought to be;
> Heavy his low-hung lip did oft appear,
> Deprest by weight of musing Phantasy;
> Profound his forehead was . . .

When age had deeply altered both men, and Coleridge had been dead nearly two years, he was still to William

> The rapt One, of the godlike forehead,
> The heaven-eyed creature . . .

Quite simply, William in later life said, 'Coleridge was the only wonderful man I ever knew.'

The two men had met briefly at Bristol nearly two years earlier, and again, when William was on the way back from that city, in spring 1797, at Nether Stowey in Somerset, where Coleridge had moved, with wife and child, the previous winter. They knew one another chiefly as poets, and, more especially, as tyros in the writing of verse-tragedy. Sheridan, manager of Drury Lane, had just asked Coleridge to try his hand at one, which he was sketching, when William saw him in Somerset. In his turn, hearing of *The Borderers*, Coleridge had encouraged William to make the piece more practicable for the stage, and this had occupied William at Racedown during April and May. It was natural, then, when Coleridge appeared at Racedown in the first week of June, that the two should fall almost at once on their dramatic manuscripts, though Coleridge took time to read through William's last completed poem, *The Ruined Cottage*. Then, wrote Dorothy, 'after tea he repeated to us two acts and a half of his tragedy *Osorio*. The next morning William read his tragedy *The Borderers*',[2] still fatally undramatic, and given to personal and meditative passages. Even Dorothy's glow-worm appeared in a digressive interlude.

> But two nights gone,
> The darkness overtook me—wind and rain
> Beat hard upon my head—and yet I saw
> A glow-worm, through the covert of the furze,
> Shine calmly as if nothing ailed the sky:

Coleridge had intended to stay ten days at Racedown. In point of fact, he stayed twice as long. He was enchanted with the Wordsworths, and they with him. None could bear the thought of separation. On 2 July* he drove them, cross-country and along the rutted Somerset lanes, forty miles of 'execrable road' in a post-chaise, to his cottage at Nether Stowey in the Quantocks. They were introduced to Sara Coleridge, overtaxed by domesticity in a damp cottage, and Dorothy added nine-month-old Hartley Coleridge to her galaxy of favourite infants. After less than a week another new friend joined them, the Londoner and schoolfellow of Coleridge, Charles Lamb. Overcrowding—three unrelated men, three women including the servant Nanny, and one child in three small bedrooms—forced the household into

---

* The exact date seems uncertain.

intimacy. Lamb was shaken by tragic experience; less than a
year before, his much-loved sister Mary, in an attack of mania,
had murdered their mother. For a time his deliciously amusing
stammered conversation was subdued, but he was grateful for
sympathy. 'It was kind of you all', he wrote afterwards, 'to en-
dure me as you did.'

Seeing Dorothy at such close quarters, in the familiar setting
of his own cottage, Coleridge drew a discerning portrait of her
appearance and character, in a letter to Cottle, the Bristol publisher of his poems.

Wordsworth & his exquisite Sister are with me—She is a woman indeed!—in mind, I mean, & heart—for her person is such, that if you
expected to see a pretty woman, you would think her ordinary—if
you expected to find an ordinary woman, you would think her pretty!
—But her manners are simple, ardent, impressive—

> In every motion her most innocent soul
> Outbeams so brightly, that who saw would say,
> Guilt was a thing impossible in her—

Her information various—her eye watchful in minutest observation
of nature—and her taste a perfect electrometer—it bends, protrudes,
and draws in at subtlest beauties and most recondite faults.[3]

To the correct Southey, husband of Sara's sister, Coleridge
conveyed by careful choice of words that Dorothy's appeal was
not sexual. 'Miss Wordsworth is a most exquisite young woman
in her mind and heart.'[4]

This first well-known description, acute as it is about Dorothy,
also tells one a good deal about Coleridge himself. The reference to Dorothy's taste as 'a perfect electrometer'—a comparatively recent popular scientific term—underlines this. For
Coleridge regarded himself as ideally wedded to all forms of
knowledge, 'to collect materials and warm my mind with uni-
versal science'. His three most famous poems—*The Ancient
Mariner*, 'Kubla Khan', and *Christabel*—are not in a style of
verse he wrote either before or after; and all three, roughly
speaking, were written or put together within the first year
after he had been in close contact with William and Dorothy
Wordsworth. It is difficult to say exactly that William and Dorothy 'inspired' them, just as it is difficult to find terms for their
friendship. The famous 'three people but one soul'[5] was

perhaps never said by Coleridge at all, though he later wrote to
Godwin that 'tho we were three persons, it was but one God',[6]
which gives, perhaps, a different emphasis to their relationship
with one another. At about the same time, too, Coleridge wrote
that he, William, and Dorothy 'have formed a deep conviction
that all is contemptible that does not spring immediately out of
an affectionate Heart'.[7]

At all events, nearly a decade afterwards, when William was
finishing his long spiritual autobiography, later known as *The
Prelude*, he did not seem able to decide whether, as intended,
the poem was addressed to Coleridge, or should have been
attributed to the guidance of Dorothy—

> the Sister of my heart
> Who ought by rights the dearest to have been
> Conspicuous through this biographic verse.

The bond of sympathy between Dorothy and Coleridge is
impossible to estimate. It certainly cannot be reckoned in any
modern Christian-name terms; Coleridge did not know until
he stayed with the Hutchinsons two years later that Dorothy's
earlier friends called her 'Dolly'.[8] A few varied tastes in common emerge in their writing. Both had an enthusiasm for the
novels of Samuel Richardson; both took special delight in the
sight of the smoke from cottage chimneys curling up from the
valleys.[9] The group's triple sympathy—'the Concern' as they
called themselves—was virtually intangible. The main point
was that all three were young and happy, yet not without tragic
experience which heightened the poignancy and intensity of
happiness. William had, as a part of his nature, his French
adventure, Coleridge had by now learnt the dangerous tyranny
of opium; as for Dorothy, she had undoubtedly felt, more deeply
even than her own brothers, the tragedy of their orphaned state
—'how are we squandered abroad'. At all events, it is surely
impossible to use the words 'love' or 'in love' with any meaning about the three. They were all in a state of delighted wonder
with one another, a type of joyful recognition and dazed bewilderment, for which there can be no rational account.

The Wordsworths had planned to stay at Nether Stowey for
a fortnight. In fact, Dorothy never returned to Racedown.
Within two days she was copying a list of errata for Coleridge

to send to Joseph Cottle, his publisher at Bristol.[10] Brother and
sister found themselves enthralled by Coleridge's wealth of
reading and the impulsive brilliance of his speech in the homely
Devon accent. 'He talked on for ever.' Their enchantment was
heightened by the beauty of the countryside in summer weather.
Two days after the Wordsworths arrived, Sara accidentally
spilt a skillet of boiling milk and scalded Coleridge's foot, con-
fining him to the garden. The guests were left to explore the
neighbourhood on their own. 'There is everything here,' wrote
Dorothy to Mary at once; 'woods wild as fancy ever painted,
brooks clear and pebbly as in Cumberland, villages so romantic;
and William and I, in a wander by ourselves, found out a
sequestered waterfall in a dell . . .'[11] She loved 'the cottages of
Somersetshire, covered with roses and myrtle, and their small
gardens of herbs and flowers',[12] and rambled the neighbourhood with her old dream of 'happiness in a little cottage and a
passing wish that such a place might be found out'. What was
found was something entirely different.

From Nether Stowey a country road runs west along the northern edge of the Quantocks. On the right hand lies the Bristol
Channel, striped with sun and shadow; across the water the
Welsh hills come and go in the clouds. On the left, after three
miles, a narrow lane behind the Plough Inn winds into a deep,
wooded glen. There, half-hidden in fern and trees, a brook leaps
down a chain of waterfalls towards the sea. During their walk on
4 July 1797, Dorothy and William took this lane past the waterfall, and went on to explore further.[13] Higher upstream from the
waterfall which so delighted them lay an old stone dog pound
and park gates. Within, a drive leads through a deer park thickly
wooded with oak, beech trees, and holly, still a good shelter in
a sudden hailstorm. After about a mile, one end of a long white
Queen Anne house appears over a shoulder of hill. 'William
and I had rambled as far as this house,' wrote Dorothy, again
to Mary. 'The front of the house is to the south, but it is screened
from the sun by a high hill, which rises immediately from it.
This hill is beautiful, scattered irregularly and abundantly with
trees and topped with fern which spreads a considerable way.'
From the north of the house there was a view to the sea, 'over
a woody meadow-country'. Wherever they turned there were
'woods, smooth downs and valleys with small brooks running

down them through green meadows. . . . The hills that cradle
these valleys are either covered with fern and bilberries or oak
woods which are cut for charcoal.'[14]

Alfoxton House stood empty, 'with furniture enough for a
dozen families like ours,' wrote Dorothy, again to Mary. The
walled garden was well stocked with fruit and vegetables, the
moss roses 'in full beauty', when they first saw it. At the park
gates the Quantock hamlet of Holford, with about a hundred
people,* straggled down its combe. More important, three miles
away at Stowey was their spellbinding new friend, anxious to
see them '*settled*' nearby. The prosperous tanner of Nether
Stowey, Thomas Poole, solid, snuff-taking, and countrified, yet
bookish and an 'extreme Democrat', had found Coleridge a cottage, and now acted for the Wordsworths. Alfoxton House was
owned by a minor, whose grandmother, Mrs St. Albyn, had a
life interest in the property. Kindly Thomas Poole 'strongly
recommended Mr. Wordsworth as a tenant', and witnessed the
agreement signed by Wordsworth and the agent John Bartholo-
mew. It was less than a fortnight since they had first come into
Somerset, and on the face of it an impulsive move. Alfoxton was
even larger than rent-free Racedown, with three parlours, nine
bedrooms,[15] kitchens, attics, cellars, and stables with clock
cupola. The rent was high for the times, at £23 a year. Cowper
had considered a house at £20 a year expensive for his cousin,
Lady Hesketh. Yet the Wordsworths apparently did not hesi-
tate; 'We heard that this house was to let, applied for it and
took it. Our principal inducement was Coleridge's society,'
wrote Dorothy, as though that were explanation enough. Their
acute money worries persisted, yet these were as nothing to
their overwhelming need to be near Coleridge. Later William
drove back to Racedown to fetch loyal Peggy, small Basil, and
the books and clothes which they had left behind, expecting to
return. The impromptu move was complete. So, swiftly, almost
casually, began one of the great ventures of English Romantic
literature.

The Wordsworths' house-warming party, on Sunday 23 July
1797, was as unconsidered as their move. On 17 July John Thelwall, 'the agitator', arrived unannounced at Stowey, having

---

* One hundred and thirteen at the census of 1801.

walked from London. He had been arrested under the 1794
Treasonable Practices Act, of which Richard Wordsworth
warned William, but had been acquitted at his trial. For the
last year he had been in correspondence with Coleridge as a fellow radical, and now hoped to find a refuge for his family near
Stowey. Coleridge was soon to warn him that 'riots . . . might
be the consequence', if he tried to settle in the district. Yet now
the idea apparently arose of a dinner party at Alfoxton House,
with Thelwall as guest of honour. This would have been
beyond the Wordsworths' means, but Poole's old mother kindly
gave a fore-quarter of lamb, which was sent over by messenger-
boy for Dorothy to cook. Fourteen people in all were expected:
Thelwall, Coleridge, Poole, and members of his progressive
circle, a formidable task for the hostess. Radical Thelwall
noted approvingly that Dorothy managed 'without any servant,
male or female', although an old woman sometimes lent a hand
and on this day another Alfoxton cottager, Thomas Jones, was
pressed into service as a waiter. The enthusiastic group arrived
about eleven in the morning at Coleridge's request, 'that we
may have Wordsworth's tragedy read under the trees'. In the
afternoon, the company sat down in the large formal dining-
room at the back of the house, where three long sash-windows
looked out to sea. Dinner was served and Thelwall evidently
felt the occasion called for a political speech. Jones, running up
and down stairs from the basement kitchen with the dishes,
described 'a little Stout Man with dark cropt Hair . . . who
after dinner got up and talked so loud and was in such a passion' that the amateur waiter, who had never in his country life
heard anything like it before, felt frightened. Thelwall himself,
the guest of honour, felt the whole Alfoxton group was 'a most
philosophical party'. It is not recorded whether women were
invited or whether Dorothy herself sat down with the guests.
Roasting at an open fire was greasy, sooty work. 'If any scrap-
ing is required', said a recipe book tersely, 'use your nails.'[16]
As sole cook, she may well have been confined to the kitchen
throughout.

At all events, she was enchanted by Alfoxton. 'Here we are
in a large mansion in a large park, with seventy head of deer
around us,' she wrote to Mary, as one of the family, a month
after their arrival. She had never liked an artificial landscape

garden, 'ruins, hermitages, etc. etc.'; here, with the sheep and
deer browsing quietly over the hillside, 'we have a living prospect'. She always remembered their walks in the park, 'wading
almost up to the knees in fern . . . and the slim fawns that we
used to startle from their couching places among the fern at
the top of the hill'.[17] The hill which 'screened from the sun'
was a domestic asset; all eighteenth-century housekeepers suspected sunlight, foe to curtains, carpets, and complexions alike.
Within the house, there was an elegant hexagonal drawing-room
to the left of the hall, but her 'favourite parlour as at Racedown'
was the further of two panelled rooms on the right, overlooking the garden. There she sat at the window, writing: 'exactly
opposite the window where I now sit is an immense wood,
whose round top from this point has exactly the appearance of
a mighty dome. In some parts of this wood there is an under-
grove of hollies which are now very beautiful.'[18] The landscape
still bears witness to the faithfulness of her description.

Exhilarated by natural beauty, Coleridge's conversation,
and William's ardent response, Dorothy seemed quite unaware
of the impression she and William made on the neighbourhood.
She went to the village, to the baker and shoemaker, or to the
farm in the combe to fetch eggs quite happily; yet, almost from
the first, William had roused local speculation, particularly by
his nocturnal walks in the direction of the coast and the Bristol
Channel. This was not just a matter of rural gossip. Since the
beginning of this year, 1797, England had lived under fear of
invasion. Former allies, Spain and Holland, had become enemies, and the Dutch fleet was ready to convey a French invasion
force. Everywhere near the coast, farmers had withdrawn their
money from banks, and hidden it, while the English Navy had
been temporarily disrupted by mutinies at the Nore and Spit-
head. Discounting the theory that William was a smuggler (a
familiar local industry), opinion reported him as 'one of the
fraternity' of English radicals, and, even more damagingly, as
a 'French jacobin for he is so silent and dark, that nobody ever
heard him say one word about politics'.[19] Incidentally, this
remark has misled biographers into saying that William and
Dorothy were thought to be French because of their dark complexions. 'Dark' here obviously means 'secretive'. Nor was
Dorothy noticed as being particularly tanned until ten years

later, after continual exposure among the Lakeland mountains.

The rumour that they were 'French people', however, had serious consequences. It was picked up, on a visit, by a former Alfoxton servant, called Mogg, who now lived at Hungerford. On his way home, Mogg talked to the former Alfoxton cook, who now worked for a Dr Lysons at Bath. Lysons took the matter seriously enough to write on 8 and 11 August to the Home Secretary about 'an emigrant family, who have contrived to get possession of a Mansion House at Alfoxton'. The man, with 'only a woman who passes for his Sister', roamed the country by night and day with his visitors, carrying camp-stools and a portfolio. All were 'very attentive to the river near them'— Coleridge was planning a poem called 'The Brook', while Dorothy and William tried to help him trace the local stream from its Quantock source to the Bristol Channel just beyond Kilve.

The Home Office at once sent a detective, named Walsh, to investigate. He first questioned Mogg at Hungerford, and formed the opinion that he was 'by no means the most intelligent Man in the world'. However, Walsh was ordered to proceed to Somerset, find out the names of the persons involved, and report back. From the Globe Inn, Nether Stowey, Walsh wrote, 'I think this will turn out no French affair but a mischiefuous gang of disaffected Englishmen.' He also questioned Thomas Jones about the Alfoxton dinner, was able to identify Thelwall as the chief guest, and concluded, 'the inhabitants of Alfoxton House are a Sett of violent Democrats'. Walsh was no fool. He was quick to remember the name 'Wordsworth' as being already known to a colleague in the Home Office. This fact, showing that William's former activities had put him on the Government files, is not usually remarked.[20] The investigation was dropped; but it was not the joke Coleridge tried to make of it twenty years later, when he pictured Walsh listening to the two poets talking about Spinoza, and suspecting that 'Spy Nosy' referred to himself and his alleged red nose.[21]

What grew beyond a joke was that Mrs St. Albyn, from whom William held Alfoxton House on a one-year lease, became alarmed at stories about her tenants, and wanted to turn them out. The stories can be guessed from a conciliatory letter which Poole, on 16 September, wrote to her, and which shows that two of the main objections concerned Dorothy. First, there was the

entertainment of the notorious Thelwall. Poole went out of his
way to explain that Thelwall had turned up in Stowey unasked
by anyone, and that his invitation to Dorothy's dinner at
Alfoxton was a matter of 'the common duties of hospitality'.[22]
Secondly—as Jones told Mogg, who in turn told Walsh—
Dorothy was observed 'washing and Mending their cloaths all
Sunday'. Poole tried to counter this by producing Dorothy's
relationship to Uncle William Cookson, whom a former vicar
of Stowey, also a Canon of Windsor, knew well. Poole himself
apparently knew enough about Dorothy's past to stress her life
at Forncett, where she 'has principally lived with her uncle',[23]
perhaps to suggest Sunday observance—not very convincingly,
since there is no hint that the Wordsworths, either at Alfoxton
or Racedown, went to church. Coleridge and the Wordsworths
were warned of Mrs St. Albyn's hostility by 'a friendly medi-
um'—probably Poole himself. Henceforth, Dorothy and William knew their tenancy of this ideal place was threatened.

Yet any such forebodings were brushed aside in the exhilaration of the approaching autumn, when William and Dorothy
explored the Quantock countryside farther and farther afield
under Coleridge's inspiration. Even the landscape of hill and
coast was transformed by the alchemy of Coleridge's magical
words and ideas. 'He talked above singing,' wrote Hazlitt,
recollecting 'his echoing voice in the woods of All-Foxden'; he
transfigured everything they saw. 'He could go on in the most
delightful explanatory way over hill and dale, a summer's day,
and convert a landscape into a didactic poem or a Pindaric
ode.'[24] Dorothy saw everything through his eyes, even when
his ideas were mistaken. It was almost certainly Coleridge,
whose notebooks about this time show an interest in the legends
of Glastonbury,[25] who persuaded her that from the park, fifteen yards above the house, she could see Glastonbury Tor.[26]
This 'prospect' she admired was, in fact, the much-closer hill of
Brent Knoll. Not that Dorothy was credulous. When the rich
bachelor Tom Wedgwood came to stay for five days in September, and produced his impractical scheme for the education
of infant geniuses, in which he wished to interest William,
Dorothy seems to have known he was talking nonsense. Nor
would she allow even Coleridge to overcome her sense of fair-
ness and justice. When he showed her, 'in full expectation of

gaining a laugh of applause', a sarcastic review he had written, she made a remark which apparently affected him for the rest of his life,[27] and checked any desire to write further reviews.

Meanwhile, their walks westward, toward the Devon border and beyond, provided occasions for some historic events in literature. Early in October,[28] Coleridge, slipping off without his companions, came back with a 'fragment . . . composed in a sort of Reverie brought on by two grains of Opium'. Dorothy's sole recorded mention of 'Kubla Khan' was to call every *can* she used for domestic water by the term *Kubla*;[29] it is possible that Coleridge, like Keats with 'La Belle Dame' a generation later, initially treated this poem of the subconscious as a subject for joking. More deliberate purpose, though less success, showed on a longer expedition all three undertook a month later. After perhaps staying at Porlock, they took the coastal track between the north heights of Exmoor and the Bristol Channel. 'Our road lay through wood, rising almost perpendicularly from the sea, with views of the opposite mountains of Wales.'[30] Just after climbing out of the trees 'to a barren top, like a monk's shaven crown'[31]—the Sugarloaf—they rejoined the road, and reached Lynmouth in the early November twilight. Next morning, a local guide showed them 'a valley at the top of one of those immense hills which open at each end to the sea, and is from its rocky appearance called the Valley of Stones',[32] west of Linton.

Coleridge proposed that he and William should collaborate in writing a lurid narrative in the manner of Gessner's *Death of Abel*, a German curiosity composed 'in a kind of loose poetry', and popularly translated. Coleridge sketched a synopsis 'as a narrative addressed by Cain to his wife', using the local colour of the Linton valley, and much more exotic stuff—Cain 'hears the screams of a woman and children surrounded by tigers', and sacrificial drops of blood are shed. Coleridge dashed off his share, but found William with a 'look of humourous despondency fixed on his almost blank sheet'. It eventually contained two rough stanzas, beginning 'Away, away', not like anything else William ever wrote, in which, however, the tigers and drops of blood were faithfully represented. 'The whole scheme', wrote Coleridge, 'broke up in a laugh', adding, with incredible casualness, 'and the Ancient Mariner was written instead'.[33]

The beginnings of *The Ancient Mariner* were, in fact, just as casual as that. Dorothy and William returned to Alfoxton to find a financial crisis. Basil Montagu had appeared in the neighbourhood. Having both lost a Chancery suit and feeling himself unable to practise law, he could pay neither his son's lodging fees nor any of the other money he owed the Words-worths. Dorothy said with no hesitation that she would continue to look after the small Basil, even at a loss; but at the same time there arrived a letter from 'Aunt Wordsworth'—Richard Wordsworth's widow—about the money still owed for William's Cambridge education, 'which Wm. has been utterly unable to answer'.[34] Coleridge too owed money all round. 'As our united funds were very small' (so William afterwards phrased it), they decided again to try a joint work, with the idea of selling it to a new periodical, the *Monthly Magazine*. On a dark, cloudy morning on 13 November, William, Coleridge, and Dorothy set out to walk eight miles to Watchet, the nearby little sea-port. Dorothy wrote laconically to Mary of 'William and Coleridge employing themselves in laying the plan of a ballad, to be published'.[35] The plan was 'laid' and the ballad begun on the same 'memorable evening',[36] when all three stayed somewhere in the jumble of little cottages and sailors' inns near the outflow of the Washford River, which formed the harbour protected by its jetty. Wordsworth's contribution to the scheme and ideas of *The Ancient Mariner* was considerable, though not to its actual composition. He himself only remembered composing the lines[37]

> And listens like a three years' child
> The Mariner hath his will

While Coleridge recollected that William was responsible for

> And thou art long and lank and brown
> As is the ribb'd sea-sand.

Though many other sources have been convincingly suggested, it is possible Dorothy too gave Coleridge one of her own favourite expressions, 'green as an emerald' (probably an echo of some folk-song),[38] early in the poem begun that night. Coleridge described 'the cold Country toward the South Pole'—

> And Ice mast-high came floating by
> As green as Emerauld.

The campaign to raise money through literature continued. In December 1797 Dorothy Wordsworth had the unexpected experience of a stay in London which arose, like so many things, from Coleridge's ebullient enthusiasm. Gloomy about the fate of his own tragedy—'emigrated to the Kichen or Clŏaca' —he recommended William's *The Borderers*, through an actor, to the management of Covent Garden; he claimed that if accepted they would 'put it in preparation without an hour's delay'.[39] Dorothy tried valiantly at first to keep her head at this exciting prospect, telling Mary she had 'not the faintest expectation' of success; yet she could not help revealing to cousin Elizabeth Threlkeld at Halifax her dreams that 'it would have a prodigious run', followed by 'wildish' schemes for spending the much-needed money it would earn.[40]

The first week of December found the Wordsworths staying with the Nicholson family, relatives of Halifax friends, at 15 Cateaton Street, while William 'altered his play at the suggestion of the principal Actors'. Then they awaited the management's decision, helped by free tickets for Covent Garden and Drury Lane. Dorothy remembered the glittering theatre, the vast space, 'men seen at the extremity of the opera house, diminished into puppets'. She wrote to Christopher, now a junior Fellow at Trinity College, Cambridge, that she had twice seen Mrs Siddons act: on Saturday, 2 December as Portia in *The Merchant of Venice*, and on Monday, 4 December in *Isabella*, an adaptation of Southerne's *The Fatal Marriage*. These performances drew a deep response from her ardent nature, and reading through a Shakespeare play later appears as one of her favourite evening entertainments. Even a ballet pantomime stirred her imagination and lingered in the memory. Years later, seeing the interlaced beams of firelight in a Highland hut, she thought spontaneously 'what a feast it would be for a London pantomime maker, could he but transport it to Drury Lane.'[41]

After all her hopes—she had even bought a new pair of shoes —by 13 December Wordsworth's play, like Coleridge's, was rejected. Dorothy in a 'very entertaining letter' to Halifax loyally maintained that she was in no way disappointed; William expected a reform in the stage when the play might appear to great advantage. Cousin Elizabeth, who privately considered

William an 'excentric' young man, commented drily that she would not give implicit credit to this assertion. 'However', she added, 'they are happy in having very fertile imaginings, which are a continual source of entertainment to them.' Their brief love-affair with the theatre over, brother and sister took the Bristol coach from the White Horse, Piccadilly, on 15 December. They spent about three weeks, including Christmas Day, Dorothy's twenty-sixth birthday, in Bristol, and were home at Alfoxton in the first week of 1798.

There they found Basil Montagu, who had been looking after his own son for the past month, and despatched him on his way to London, with a batch of new-made shirts for their brother Richard. Coleridge greeted them with three pieces of news. First, he had accepted an invitation to be a Unitarian preacher at Shrewsbury. Second, he had received, and had refused, an offer of £100 from Thomas Wedgwood and his brother Josiah. Thirdly, he now knew that William and Dorothy would not be allowed to stay at Alfoxton beyond 'early summer'. Coleridge left on 12 January 1798 for Shrewsbury. There he received an even more munificent offer from the Wedgwoods—an annuity of £150 for life. This he at once accepted, partly to relieve his mountain of debts, but also, as he wrote to William and Dorothy, to afford to be with them frequently when they left Alfoxton, so 'that wherever your after-residence may be, it is probable that you will be within the reach of my Tether'.

Faced with the certainty of leaving Alfoxton within a few months, Dorothy had one resource. She began to keep a journal of her life there.

7. *'Lyrical Ballads'*

It may, of course, be a coincidence that Dorothy started her Journal on 20 January 1798, shortly after the first news that she was to leave Alfoxton; but, if so, it is a strange one. She had been there over six months already, without, apparently, writing a word. No one seems to have suggested it to her for any reason. She was not, for instance, stimulated by Coleridge's conversation for he was away, and did not return till 9 February, when she had been writing for three weeks; moreover, attempts at description in his own Notebooks are as different from hers as those of two writers can be, especially when, a few years later, they are portraying an identical scene.[1] It has been conjectured, too, that she wrote for William, 'at his suggestion, and for his special pleasure'.[2] There seems no evidence for this, though William did copy the first four sentences into his own notebook.

At all events, the Journal began:

The green paths down the hillsides are channels for streams. The young wheat is streaked by silver lines of water running between the ridges, the sheep are gathered together on the slopes. After the wet dark days, the country seems more populous. It peoples itself in the sunbeams. The garden, mimic of spring, is gay with flowers.

Though utterly fresh, the early stages of this Journal show great pains to achieve improvement. 'The country seems more populous' is at once redefined as 'It peoples itself in the sunbeams'. Conventional phrases, such as 'mimic of spring', seldom appear later. On the second day (21 January) she tries a piece of fanciful whimsy—'Moss cups more proper than acorns for fairy goblets'—in a manner she never repeats. After these few early uncertainties, her style is assured. In literature, as in anything, Dorothy was a quick learner.

The paradox of her unique style is that it is no style. The Alfoxton entries are practically in note-form. Nothing distracts. The acute observation by Dorothy is there, but no Dorothy herself. Every object, sight, sound is allowed its own nature. A warm day is built from a series of little impressions, matter-of-fact but cumulative.

Midges or small flies spinning in the sunshine; the songs of the lark and redbreast; daisies upon the turf . . . the hazels in blossom; honeysuckle budding.

Where she elaborates, every word counts.

Sat a considerable time upon the heath. Its surface restless and glittering with the motion of the scattered piles of withered grass, and the waving of the spiders' threads.

Though not merely local, the four-month journal is a record of that region, where the shaft of the Quantock ridge thrusts westward to Exmoor, never far from the sea, and subject to instant extremes of weather. The sudden blanketing of sea-fog, the equally abrupt onset of violent winds play a constant part. One entry, meticulous in its striving for exact impression, combines the two.

The sea at first obscured by vapour; that vapour afterwards slid in one mighty mass along the sea-shore; the islands and one point of land clear beyond it. The distant country (which was purple in the clear dull air), overhung by straggling clouds that sailed over it, appeared like the darker clouds, which are often seen at a great distance apparently motionless, while the nearer ones pass quickly over them, driven by the lower winds. I never saw such a union of earth, sky, and sea.

Yet Dorothy's journal entries are seldom solely prose descriptions. At best, they contain a kind of controlled poetic comment. The largest passage of the Alfoxton Journal begins prosily enough: '26th [February] Coleridge came in the morning, and Mr. and Mrs. Crewkshank'—John Cruikshank was the future tenant of Alfoxton, come to view the property. Dorothy goes on:

Walked with Coleridge nearly to Stowey after dinner. A very clear afternoon. We lay sidelong upon the turf, and gazed on the landscape till it melted into more than natural loveliness. The sea very uniform, of a pale greyish blue, only one distant bay, bright and blue as a sky; had there been a vessel sailing up it, a perfect image of delight. Walked to the top of a high hill to see a fortification. Again sat down to feed upon the prospect; a magnificent scene *curiously* spread out for even minute inspection, though so extensive that the mind is afraid to calculate its bounds. A winter prospect shows every cottage, every farm, and the forms of distant trees, such as in summer have no distinguishing mark. On our return, Jupiter and Venus before us. While the twilight still overpowered the light of the moon, we were reminded that she was shining bright above our heads, by our faint shadows going before us. We had seen her on the tops of the hills, melting into the blue sky. Poole called while we were absent.

Early in the journal, Dorothy shows her gift for describing not only sights but sounds. The kind of distinctions she makes seem to show that her hearing was abnormally acute. On 23 January, she wrote:

The sound of the sea distinctly heard on the tops of the hills, which we could never hear in summer. We attribute this partly to the bareness of the trees, but chiefly to the absence of the singing of birds, the hum of insects, that noiseless noise which lives in the summer air.

Next day, the 24th:

The half dead sound of the near sheep-bell, in the hollow of the sloping coombe, exquisitely soothing.

On the 27th, after a walk near the little industrial water-mill in the Holford valley—

The manufacturer's dog makes a strange, uncouth howl, which it continues many minutes after there is no noise near it but that of the brook. It howls at the murmur of the village stream.

Finally, she comments on the huge gale, which swept across Southern England on 1 February.

The trees almost *roared*, and the ground seemed in motion with the multitudes of dancing leaves, which made a rustling sound distinct from that of the trees.

These intense impressions of sound were gathered, it may be noted, when Coleridge was absent from her walks; his conversation, though stimulating, could obliterate all else, as Keats found twenty years later—'I heard his voice as he came towards me—I heard it as he moved away—I had heard it all the interval—if it may be called so.'[3]

Several entries use almost the same words as Coleridge and William employ in their poems, and much has been written on the relation between Dorothy's journal and the poets' work. Apart from one or two instances, though, it is impossible to say which came first, or who can be said to derive one from the other. Dorothy, indeed, often records a bare fact, which emerges in a poem. 'William wrote a description of the storm' or, more specifically, 'William wrote some lines describing a stunted thorn'. Yet where there is most likeness between a journal entry and a poem, it is usually difficult to dogmatize. On 25 January, Dorothy and William walked by night along the lower road to Stowey, to call on Poole. William composed some lines 'extempore'—he could compose dozens of lines in his head—containing an account of the moon suddenly appearing:

> There in a black-blue vault she sails along.

Dorothy's journal entry is almost exactly alike, in this and many other phrases. 'At once', she wrote, 'the clouds seemed to cleave asunder, and left her in the centre of a black-blue vault. She sailed along . . .'. Yet Coleridge, who was not present that night, and did not return to the neighbourhood for another fortnight, made in his notebook a poetical descriptive comment on a skyscape using the precise and unusual term 'black-blue', instead of the more common 'blue-black'.[4] The debt, as between the three friends, is practically impossible to determine. It can only be said that in literature as well as life, this was a time of the most intensely shared sympathy, which even the disasters of the future could never wholly erase.

At some time, though, in this otherwise happy spring, Dorothy appeared shaken by an unfortunate incident connected with Coleridge, the details of which will perhaps always remain obscure. In the previous October, Coleridge's book of poems had been reprinted by his Bristol publisher, Joseph Cottle. To this second edition were added some poems by Charles Lamb and Charles Lloyd, Coleridge's protegé, who had briefly visited Alfoxton that September. In November, Coleridge, with his fatal impulse to be 'devilish clever', published in the *Monthly Magazine*, under an assumed name, 'three sonnets in ridicule of my own, and Charles Lloyd's, and Lamb's'. Lamb was naturally offended, and so was Coleridge's brother-in-law, Southey, who mistakenly thought himself parodied. Lloyd, mentally unstable, was temporarily living with Southey in London and saw Dorothy when she and William called in December 1797. In spring 1798, Lloyd is said to have written her a letter, calling

Coleridge a villain, and 'proving that this was her opinion no less than his'; she, in turn, is said to have shown the letter to Coleridge 'with tears'.[5] It has to be remarked that the only authority for this, so far as it concerns Dorothy, is a note written by Coleridge twelve years later, in the throes of a morbid suspicion, produced by his addiction, of all his friends. Dorothy herself, reading in April 1798 Lloyd's novel, which in turn satirized Coleridge, merely commented that it 'bears the marks of a too hasty composition'.[6] Her agitation may be shown in some journal entries, 29-31 March, which seem strangely brief and out of character: 'Coleridge dined with us'; 'Walked I know not where'; 'Walked'.

On the other hand, with the loss of Alfoxton now certain, March was an intense time. In the spring of 1798 Martha Ferguson wrote, 'Dorothy, from whom we have heard lately, is deeper in plays and poetry than ever.' This is certainly confirmed by a letter of 5 March, in which Dorothy copied for Mary Hutchinson some 375 lines of *The Ruined Cottage*, which had 'grown' under Coleridge's influence, cautioning her not to 'let this poem go out of your own hands'. Every copy, representing hours of work, was precious. In return, Mary would send by wagon a ham and a cheese from her brother's farm, for the Wordsworth household. The letter, ending 'God bless you dear Mary William's very best Love', is as intimate in tone as any Dorothy wrote to her brothers, and establishes Mary's place as a member of the family circle. As such, she was admitted to their plans about a future home when the Alfoxton lease ran out. On the next day, William confirmed to a Bristol acquaintance, 'We leave Alfoxton at Midsummer. . . . I am at present utterly unable to say where we shall be.'[7] The chief concern of both was still Coleridge's society, 'so important an object', wrote Dorothy, 'that we have it much at heart'. She hoped that they might find another house near Stowey.[8]

Suddenly, within a week and without warning, the whole picture changed. Once again, the mercurial temperament and boundless intellectual curiosity of Coleridge transformed the scene. He needed, for 'my intellectual utility', to read the German theologians and philosophers, above all Kant.[9] Then a bolder plan characteristically formed in his mind. Between 6 and 11 March 1798 a 'delightful scheme' was hatched: the Col-

eridges, Wordsworths, and friends would form 'a little colony'
in Germany·to learn the language and study 'natural science'.
It should be in a village, of course, near a university, in moun-
tainous country and near Hamburg to save travel expenses,[10]
wrote William, though German geography unfortunately
proved disobliging over the last matter. 'Our present plan is to
go into Germany for a couple of years,' wrote Dorothy to
Richard,[11] whose response to this surprising news can only be
imagined. The future settled, they were able to enjoy the un-
folding beauty of the countryside, and turn their minds to
poetry.

The only way to raise money for the expenses of the journey
to Germany would be for William to sell his poems to a publisher. On 23 March Coleridge read the completed *Ancient
Mariner* to William and Dorothy; tradition says that this took
place in the panelled parlour at Alfoxton House. William at this
time spoke little of his own poetry except in private conversation with Coleridge;[12] but Dorothy noticed that his powers
seemed to expand every day and 'his ideas flow faster than he
can express them'. By 13 April 1798 William wrote to Joseph
Cottle, who had already heard some new work on a visit to
Stowey, that he had added to his 'stock of poetry'. He invited
Cottle to stay at Alfoxton to hear more poems under the old
trees in the park. On 9 May the invitation was repeated with
warm messages from Dorothy, who liked Cottle; 'we will never
forgive you if you do not come'.[13] Publication in some form
seems to have been taken for granted by everyone. Already
Dorothy had walked over to Stowey with William 'to have his
picture taken' by the visiting artist, W. Shuter,[14] the first known
portrait of the poet, which Coleridge gave to Cottle. She and
Coleridge walked with William to Cheddar, on his way to
Bristol.

So it was that one day, during their last full month at Alfoxton, found Dorothy alone at home, while William consulted
with his publishers at Bristol. On that afternoon in late May,
two visitors appeared: Coleridge had brought his own guest,
William Hazlitt, over from Stowey to see the 'romantic old
family mansion' and Wordsworth's poems. Dorothy gave
Hazlitt free access to the manuscripts of *Lyrical Ballads* into
which, he said, he dipped 'with the faith of a novice'. She also

put both men up for the night, a gesture of spontaneous hospitality which would have been unthinkable for an unchaperoned spinster a generation later. Hazlitt kept a vivid memory of the old room with blue hangings in which he slept, walls covered with 'round-faced family portraits' of long-dead St. Albyns, the belling of stags from the wooded hill at dawn, and the fallen tree-trunk in the park on which he sat after breakfast. Dorothy, by contrast, may have proved something of a disappointment, for Hazlitt—an enthusiastic walker—had one favourite topic of conversation on a journey, 'and that is what one shall have for supper'. His only recorded memory of the poet's sister was that she 'kept house and set before us a frugal repast': probably, with William away, all she had in the larder.[15]

The first four or five months of 1798 had been, as Dorothy noted, a time of intense productivity for William: first, mainly blank verse, but then in March, April, and May, a whole succession of ballad-like lyrics, narrative or philosophic, large or small. From about 24 May Dorothy stayed with Sara Coleridge, whose second child had been born ten days earlier; meanwhile William visited Bristol and returned bringing Joseph Cottle. Dorothy's journal is non-existent for the events at Alfoxton in the last week of May, and the reader is forced to rely on Cottle's own account. He shared a vagueness about dates with a fondness for jocular detail, and his narrative, written long afterwards, may be read with reservations. He drove Wordsworth and Coleridge, he wrote, in his gig with 'philosophers' viands': a bottle of brandy, a noble loaf, and a stout piece of cheese. On the road from Stowey a sturdy beggar smelt the cheese and stole it. When they unhitched the horse in the courtyard, the bottle of brandy rolled down the tilted cart and smashed on the ground. 'We might have collected the broken fragments of glass, but the brandy! that was gone! clean gone!', leaving a smell which perhaps caused the village gossip that Wordsworth dealt in smuggled liquor. The poets showed themselves (according to Cottle) totally incompetent at unharnessing, until Peggy came out to slip off the collar and rescue the horse. They sat down finally to a dinner garnished with Dorothy's lettuce from the garden. 'At the top of the table stood a superb brown loaf. The centre dish presented a pile of true coss lettuces, and at the bottom appeared an empty plate, where the "stout piece

of cheese" ought to have stood.' The company drank spring
water, or, as Cottle inevitably called it 'Castalian champagne'.
Dorothy evidently decided not to risk masculine housekeeping
again; for their further meals, the village a mile off provided all
that a publisher or two authors could demand.[16] By the last day
of May she reported triumphantly to Richard at Staple Inn that
'William has sold his poems very advantageously—he is to
receive the money when the printing is completed'.[17]

The plan for a volume had been discussed when the poets took
Cottle on their favourite tour to Lynton, apparently leaving
Dorothy at home. After many other proposals over the past six
months, the book, *Lyrical Ballads* was to follow, more or less, an
original project of the previous November, when, according to
Dorothy,[18] Coleridge's then-unwritten *Ancient Mariner* was 'to
be published with some pieces of William's'. Now the finished
Coleridge ballad could be joined to William's new book of lyrics,
in which Cottle found 'a peculiar but decided merit'. The motive,
William later wrote, was the same as it had been earlier: 'I
published those poems for money and for money alone',[19] and
the two poets, referred to as 'The Author' in a short preface,
were to remain anonymous. The whole volume was indeed
'peculiar'. The *Mariner*, given pride of place, was accompanied
by three other poems from Coleridge, one in dramatic form.
There were four earlier-written poems, or sections of poems,
by William; but what gave the book its character were the lyric
poems of various lengths William wrote at Alfoxton, the fruit of
Dorothy's care of him, no less than her ceaseless housekeeping,
which, in one poem personally addressed to her, she is invited
to abandon for the poet's and her own benefit—'Come out and
feel the sun.'

After about a week at Alfoxton, Cottle returned to Bristol,
taking a copy of the *Ancient Mariner* and some of William's
poems. Immediately after, probably on Monday, 4 June, William followed him, for a further meeting. The date was fast
approaching to leave Alfoxton, 'that Dear and beautiful place',
and Dorothy could not completely hide her regrets. On 25 June
they went for a week to the Coleridges, then on to Bristol,
where they probably lodged with Cottle in Wine Street, the
busy heart of the city. 'You can scarcely conceive', wrote
Dorothy to her aunt Rawson on 3 July, 'how the jarring con-

trast between the sounds which are now for ever ringing in my ears and the sweet sounds of Alfoxton makes me long for the country again.' Yet from pride and loyalty to William she was determined, as with the London visit, to put a good face on things; the letter was designed to reassure. They were glad, she wrote, not to be 'shackled with the house', because the journey to Germany would equip them to translate, which she mistakenly believed to be 'the most profitable species of literary labour'. Life in Germany would be cheap, the country beautiful, and the entire Coleridge family would be with them, though this, of course, did not happen. Poor Basil and Peggy must, it was true, be left behind, after nearly three happy years together, but travel was hazardous for children and their faithful servant had in any case married. To counter the last possible doubts of an anxious aunt, Dorothy stressed that the German journey, and their life generally, was economical: 'have lived upon our income and are not a farthing poorer than when we began house-keeping'.[20] Here again, her calculations were optimistic: paying the final bills at Alfoxton—rent, servants' wages, tradesmen— she had written urgently to Richard for an extra £30.[21] The let- ter to Mrs Rawson reads ingenuously; it was herself Dorothy was reassuring, as well as her aunt. The Fergusons, as always, spoke with Yorkshire caution of the 'Plan . . . which you may think a curious one. . . . Dorothy lays it all down in a very agreeable manner—and I wish it may answer all their Views.' Even they, though, were impressed by the news that William was to publish his poems. 'I believe he has got a good price from the Bookseller,' wrote Cousin Edward to Cousin Samuel, in August, with admiration and an undertone of surprise.[22] 'Cottle', Dorothy confirmed a month later, 'has given thirty guineas for William's share of the volume.' Any doubts about the German project were finally suppressed.

The printing and production of the *Lyrical Ballads* tethered the Wordsworths to Bristol throughout July and August 1798. They found time for two brief trips into the country. One, 'a dash into Wales' on 4 August, was a week's visit with Coleridge to Thelwall at his farm in Brecknockshire. Another, earlier expedition was the occasion of a masterpiece. It came about, like many of their ventures, almost by chance. On 8 July they went to visit an acquaintance from William's Paris days, James

Losh, who was at Bath for his health. A fellow guest at dinner
had just published *A Walk through Wales*, with an aquatint of
Tintern Abbey as frontispiece,[23] and the sight of this stirred
irresistible memories. Next day the Wordsworths returned to
Bristol and the following morning took the ferry across the
Severn, crowded with coalships from the Forest of Dean.
Nothing could show more vividly the impulsive freedom of those
footloose early years. A ten-mile walk took them to Tintern
Abbey, where they slept. By William's own account, 'Next morning we walked along the river through Monmouth to Chepstow
Castle, there slept and returned the next day to Tintern, thence
to Chepstow.' Chepstow could not hold them; they took a boat
back, upstream, to sleep for one more night at Tintern, 'and
thence back in a small vessel to Bristol'. They had covered
more than fifty miles in three days. On the walk and in the
small ship, William was composing without writing down or
altering a line in his mind. On the road into Bristol he finished
the 'last 20 lines or so' of his *Lines composed a Few Miles above
Tintern Abbey*. These '20 lines or so', composed at Dorothy's
side as they walked from the Downs into the city, are both a
thanksgiving for their past year, a hopeful prophecy of future
years, and a tribute of deepest affection for his 'dear, dear
Sister!'.

                          Therefore let the moon
          Shine on thee on thy solitary walk
          And let the misty mountain winds be free
          To blow against thee: and in after years,
          When these wild ecstasies shall be matured
          Into a sober pleasure, when thy mind
          Shall be a mansion for all lovely forms,
          Thy memory be as a dwelling-place
          For all sweet sounds and harmonies; Oh! then
          If solitude, or fear, or pain, or grief
          Should be thy portion, with what healing thoughts
          Of tender joy wilt thou remember me,
          And these my exhortations! Nor, perchance,
          If I should be where I no more can hear
          Thy voice, nor catch from thy wild eyes these gleams
          Of past existence, wilt thou then forget
          That on the banks of this delightful stream
          We stood together; and that I, so long

A worshipper of Nature, hither came,
Unwearied in that service: rather say
With warmer love, oh! with far deeper zeal
Of holier love. Nor wilt thou then forget
That after many wanderings, many years
Of absence, these steep woods and lofty cliffs,
And this green pastoral landscape, were to me
More dear, both for themselves, and for thy sake.

Written down, the whole poem, immeasurably William's finest so far, formed the finale for the new book, its tragic irony concealed till four decades later in Dorothy's life. After this idyll, the sight, sound, and dirt of Bristol seemed more 'hateful' than ever. From about 15 July they moved into a cottage at Shirehampton five miles out of the city, probably sublet to them by Losh. Here they stayed while the poems were in the press. The last week in August saw them on the road to London, with tourist visits to Blenheim and Oxford on the way.[24] From London, on 13 September, Dorothy wrote that the poems were 'printed but not published, in one small volume'.[25] By publication day, the two authors were at sea, headed for Germany.

The party which left London for Yarmouth on 14 September 1798 was smaller than they had hoped. Coleridge had decided to leave his family at home; and as he wrote, 'Mrs. Coleridge's wishes tend the same way'. The only volunteer for the educational colony was a young Stowey neighbour, John Chester, as Hazlitt said, 'attracted to Coleridge's discourse as flies are to honey'. It was Dorothy Wordsworth's first experience of foreign travel: eight years since she had traced William's paths through Switzerland upon the map and copied passages from his descriptive letters for Jane's admiration. The present reality was very different. They sailed from Yarmouth on Sunday morning, 16 September, to find the North Sea its unpredictable self. 'Wordsworth shockingly ill,' reported Coleridge, 'his sister worst of all—vomiting and groaning unspeakably. And I neither sick nor giddy but gay as a lark',[26] which suggests that his habitual domestic remedy stood him in good stead. Yet in spite of the miseries, Dorothy stuck to her resolution of keeping a travel diary. By Tuesday morning they were in the still waters of the Elbe, but fog delayed them. 'The air was cold and wet, the decks streaming, the shores invisible, no hope of clear

weather.' Slowly the fog cleared, the ship nosed its way
upstream, and from Altona a rowing boat landed them by
Thursday afternoon on the quay at Hamburg.[27] William afterwards called it 'a very pleasant voyage of three days and three
nights', but this was a charitable description.

The observation in Dorothy's travel diary is sharp and vivid,
yet without the transforming joy of the Alfoxton journals; it
reads interestingly, but as though her heart were not fully in it.
She attempted to describe picturesque sights: Dutch women in
straw bonnets the size of umbrellas, German girls in white caps
with long ribbons, while at the same time her truthful eye registered 'dirty, ill-paved stinking streets' leading to their inn. Her
bedroom was plastered with a coating of dirt, 'a *man*' brought
her washing water and the landlord sat 'with greasy face at the
head of the table, laughing with landlord-like vulgarity'. Next
day she went with Chester to walk among the crowds on the
Promenade in the morning and to the French theatre in the evening: 'a mixture of dull declamation and unmeaning rant . . . the
story was carried on in singing', perhaps a spontaneous English
reaction to Continental opera. Sturdily insular, baffled by a
strange language and currency, both Wordsworths strongly
resented the usual sharp practice of a large seaport. Worsted in
argument by a baker, and insulted by an 'insolent' porter, William carried his own trunks through the rain. 'I have constantly
observed a disposition to cheat, and take advantage of our
ignorance,' Dorothy wrote indignantly.

The high point of their stay in Hamburg came on 26 September when the party dined with Herr Klopstock, a merchant to
whom Coleridge had an introduction, to meet his famous
brother. The father of German poetry was homely and cheerful
despite his swollen legs and his seventy-three years. Revering
this great poet, but maintaining her usual honesty, Dorothy did
not pretend to any literary conversation—'his French is even
worse than mine'—but her observation of the scene was vivid:
the little girl of seven who sat up to dine, the servant who laughed
and talked freely during the meal, and the dress: 'N. B. Klopstock the poet's lady much exposed'. The five-course dinner
was heavy and two days later she still had a headache.[28] The
two poets set out to find permanent lodgings, but Coleridge,
drawing on the Wedgwood annuity, could be far freer in his

choice. On 30 September Coleridge and Chester left for Ratze-
burg, a charming island-resort on a lake about thirty miles out
of Hamburg, 'all in high life among Barons counts and count-
esses', where, as Dorothy later said, 'we should have been
ruined'. They explored Altona and the pretty villages along the
river bank, with gravel walks among the fields, gardens, neat
houses, and 'window curtains white as snow': all much too expensive for them to rent. They must find 'obscurer and cheaper
lodgings without boarding'.[29] To separate was wise if they
meant to learn German, but the parting from Coleridge did not
go unremarked among their friends. 'I hear that the two noble
Englishmen have parted no sooner than they set foot on German
earth,' wrote Lamb sardonically to Southey.[30] Without Coleridge's company, Hamburg seemed increasingly oppressive.
The city gates were shut at half-past six; the citizens emptied
their bowels in the streets, which stank horribly. Dorothy saw
a man beating a middle-aged woman in public, and a surly-
looking man 'driving a poor Jew' with curses and blows from
a stick, while bystanders looked on with 'cold, unfeeling
cruelty'.[31] 'It is', as William wrote to Poole, 'a *sad* place.' He
asked his friend to see if, by any chance, they might recover
Alfoxton, but nothing came of this.

In preparation for a journey, William called at Klopstock's
to enquire the way into Saxony and at Remnant's English
bookship to buy Burger's *Poems* and Percy's *Reliques of English
Poetry*. At five in the evening on Wednesday, 3 October, they
took the diligence for Brunswick. English travellers had no idea
of the carts and 'dreadful roads' in the German provinces;
Dorothy fell ill from the jolting with 'pain in my bowels' followed
by sickness.[32] By Friday evening they reached Brunswick,
where they slept at an inn. The last stage, twenty-five miles
from Brunswick to Goslar in Lower Saxony, lasted from eight
in the morning until eight at night. They had intended to tour
around, but it was four months before they could face travelling
again.

Goslar, on paper, was an attractive choice: near the Harz
mountains, a medieval 'free city' with its own laws until 1802,
the Romanesque palace and cathedral built by the Emperor
Henry III before 1050. In real life it had been sacked by the
Swedes in the Thirty Years' War and wrecked by two disastrous

fires in the eighteenth century. It had become a decaying small town; 'once the residence of Emperors', as William wrote, 'now the residence of Grocers and Linen-drapers'.[33] Even Dorothy, always inclined to make the best of things, admitted 'it is a lifeless town'. They could not afford to board with a professional family, but settled into cheap lodgings with a draper's widow, Frau Deppermann in Breitstrasse. Dorothy wrote only one letter to Halifax. 'I think', observed shrewd Cousin Martha, 'it does not altogether answer their expectations.'[34] In fact nothing could have been less like the literary circle they imagined in the heady days at Alfoxton. 'Provisions very cheap and lodgings very cheap,' they both reported to Coleridge after five weeks, 'but no Society.'[35] They sat alone together at mealtimes and in the evening by the black iron stove. Conversation was restricted to their landlady, a French *émigré* priest, a young apprentice in the house, and a willing but totally deaf and toothless neighbour. They received no invitations. Dorothy believed this was because they could not afford to offer hospitality, but Coleridge in private put the matter more bluntly: only married women went into society. 'His taking his Sister with him was a wrong Step. . . . Sister here is considered only a name for Mistress.'[36] Living as they did, Dorothy hopefully reported 'tolerably regular progress' at German, 'but if we had had the advantage of good society we should have done much more'. William seemed baffled that he had learnt 'more french in two months than I should acquire German in five years'; had he forgotten the charm of Annette's tuition? 'William works hard', wrote Dorothy, 'but not very much at the German.'

   In December, the iron cold of a central European winter closed in on them; 'the cold of Christmas day has not been equalled even in this climate in the last century,' Dorothy later wrote to Christopher. In fact there was hard frost with driving snow over Northern Europe generally, 'as severe as any ever felt in England,' wrote Parson Woodforde on 9 February 1799. In Goslar, Dorothy suffered from 'air that sets the flesh a creeping, even when you go through the passages and staircases', and put on greatcoats to shift from room to room. They wanted to move, but she was 'afraid of travelling all night in an open cart' on the dreadful roads.[37] They walked in the snow, muffled in furs, for one hour each day. Otherwise she perfected at close quarters the skill she had learnt in England, to subdue herself utterly to the demands of William's poetry. For, having no books, he felt 'obliged to write in self-defense'. For inspiration he went 'walking by moonlight in his fur gown and black fur cap in which', wrote Dorothy playfully, 'he looks like any grand Signior', perhaps resembling the actors she had seen just a year previously in *The Merchant of Venice*.

The poems included, William wrote to Coleridge, 'two or three little Rhyme poems which I hope will amuse you'—that is, 'interest you'*—in return for some lines which were meant to 'amuse' in a more modern sense, Coleridge's hexameters[38] addressed to 'dear William and dear Dorothy', with the line, 'Dorothy, eager of soul, my most affectionate sister'. Were the versions of the two 'Lucy' poems Dorothy now copied addressed by William to her also? One, which later became the poem beginning 'Strange fits of passion', was certainly 'a favourite' of Dorothy's, but this does not mean it was addressed to her; nor was, necessarily, the other, 'She dwelt among the untrodden ways'. William followed these two with a third, perhaps the best-known of the 'Lucy' cycle.

> A slumber did my spirit seal;
> I had no human fears:
> She seemed a thing that could not feel
> The touch of earthly years.
>
> No motion has she now, no force;
> She neither hears nor sees;
> Rolled round in earth's diurnal course
> With rocks, and stones, and trees.

Coleridge thought that this 'sublime epitaph' was composed by William because 'in some gloomier moment he had fancied the moment when his sister might die'. Yet neither these, nor the later 'Three years she grew in sun and shower', nor 'I travelled among unknown men' can be exclusively associated with Dorothy. The later and similar 'Glow-worm' poem, which definitely describes an incident between her and William at Racedown, was at first addressed to 'Emma', his usual poetic name for her. 'Lucy'—always dead in the poems of this group

---

* Many biographers misinterpret this eighteenth-century usage.

—may just possibly recall the early death of Margaret Hutchinson, and the sorrow it had brought both her sister Mary and Dorothy, but that is all that can be said.[39]

By early 1799 the Wordsworths admitted that Germany was much more expensive than they had expected; William felt he did not know '*anything* of the German language', and they agreed to go home with the first fine weather of spring.[40] Early in February Dorothy wrote to Christopher that they had sent off their luggage and hoped to 'make a little circuit from town to town' on their way towards Hamburg. Three weeks later, the sun shone on the garden and at one in the afternoon they set off on foot through the skirts of the Harz Forest, where the brilliant green moss under the trees made Dorothy's eyes ache after weeks of snow. They planned a two months' ramble, largely in Thuringia, visiting Weimar—though they did not mention Goethe—Erfurt, and Eisenach, where Luther had made his translation of the Bible into German. They kept a joint journal, which, in 1800, Coleridge hoped to incorporate in a book on Germany for Longmans; but the book did not appear and the journal of their journey was lost.[41] On 22 February they were at Nordhausen, where they found 'long desired letters' from Coleridge who had moved to Göttingen to be near the University. Dorothy burst open the seals impatiently, they 'devoured' the news, and five days later sat down to write a long joint letter in reply. Dorothy recorded the joy of being out of doors, 'with the song of the lark, a pair of larks a sweet liquid and heavenly melody heard for the first time after so long and severe a winter'. She took the hardships in good part, walking 'above the ankles in water and sometimes as high in clay' and sleeping at an inn 'in company with our host and hostess and four children, a facetious shoe-maker, a Prussian tax gather[er] and a journeyman hat-maker'. What she found hardest was to be 'stared out of countenance' by curious crowds, avid for the strange sight of a foreign woman.[42] She thought it wrong to judge national character from chance examples, but in each town they came to the inquisitive crowd seemed 'barbarous', 'impudent and vulgar', even 'vicious'. Try as she might, she was homesick for England.

On about 20 April they reached Göttingen, in the southern territories of Hanover, and after six months' separation saw Coleridge again; it was an emotional meeting, for they had missed each other deeply. Dorothy could almost have kissed the packet of Coleridge's letters, while now William 'was affected to tears at the thought of not being with me'.[43] Dorothy grieved over 'Poor Berkeley', the baby whose birth she had witnessed and who had died at ten months of tuberculosis in the damp cottage at Stowey while she was far away in freezing Goslar. They spent a day together, and Coleridge walked five miles on their way with them, to delay the moment of parting. He found both Wordsworths 'melancholy and hypp'd'. and burning with 'impatience to return to their native country', in which he shared.[44] Sometimes he felt '*a huge hankering for Alfoxton*', and Dorothy imagined them all together 'in the North of England amongst the mountains whither we wish to decoy you'.[45] Whenever they made plans for the future it was for a shared life, with the casual visits, the walks and talks, the constant inspiration of the *Lyrical Ballads* days. Many years were to pass before this vision faded.

Meanwhile they made for Mary Hutchinson, the faithful friend who would give them shelter, since once again they had no home. They travelled as fast as they could by diligence to Hamburg, river-boat to Cuxhaven, and a smooth passage of two days to Yarmouth. On 1 May 1799 they landed and went straight to the North. The big old farmhouse, Sockburn, enisled by the Tees, already sheltered Mary's two younger brothers, Tom the farmer and George, and her sisters Sara and Joanna, all of whom took the Wordsworths into their family circle. John Hutchinson lived in Stockton and Henry was at sea. From May to December 1799 Dorothy and William lived with the Hutchinsons on their farm, while 'still quite undetermined where we shall reside'. They thought of taking a house nearby; 'how they propose to add to their income I cannot tell,' remarked Aunt Rawson drily. Outwardly little happened. Dorothy rejoiced that John returned safely from his trading voyage to Bengal in the *Duke of Montrose*.[46] She wept, with real feeling, over the death of Uncle Crackanthorpe who had been so unaccountably fond of her, and found herself the only one of her family mentioned in his will with a legacy of a hundred pounds: little enough after all they had lost, though 'I daresay he would have done much more if he had been a free agent'.[47] Yet this uneventful time had its inner momentum. The months after the visit to Germany confirmed a change in the political beliefs of Wordsworth and Coleridge which would alter the whole mental climate of Dorothy's future life.

'We are right glad to find ourselves in England,' wrote William three weeks after their return, 'for we have learnt to know its value.'[48] Six years earlier he had returned from France, attacking the English government's war against the French Republic as a 'giving up to the sword' of the defenceless poor. His pamphlet 'by a Republican' had been considered by the leading radical bookseller, Johnson of St. Paul's Churchyard, too dangerous for publication. Even the wounds of the Terror he had hoped would be healed by the death of Robespierre, and society 'march firmly towards righteousness and peace'. He remained hopeful when General Bonaparte led the republican armies into Holland, Belgium, and Germany west of the Rhine; he studied Machiavelli in John Pinney's library at Racedown to analyse the contradictions of his own times.[49] With the brilliant campaigns across the Alps in 1797, he had even wondered briefly whether Napoleon might prove the creator of a united Italy. Disillusionment began in January 1798 with French attacks on Switzerland, Coleridge's 'shrine of Liberty'. The new Helvetian Republic was clearly a puppet of France. The two friends' mistrust increased as Napoleon's personal power increased; by 10 September 1799, four months after their return from Germany, Coleridge hoped William would write for 'those who in consequence of the complete failure of the French Revolution, have thrown up all hopes of the amelioration of mankind'.[50] By 9 November their worst suspicions were confirmed, when after the *coup d'état* of Brumaire Napoleon became First Consul, with sweeping personal powers. So started a long and painful loss of faith, in which Britain, resisting the French tyrant, began to appear gradually as the defender of liberty. Wordsworth and Coleridge had a long conversation in March 1801, in which they 'deeply regretted' any writings of what a correspondent called 'jacobinical pathos'. Those who wrote to stir up revenge for human suffering, they agreed, 'are bad poets and misguided men'.[51] Dorothy Wordsworth professed no independent opinions on the history of her own times. She had eagerly shared her brother's

revolutionary hopes and now shared his growing disillusionment. She slowly developed a hatred of Napoleon, a mistrust of
the French, except, characteristically, of 'poor Annette', and
eventually a growing suspicion of liberal activity at home. During these quiet months the scene was set for brother and sister
to harden into the marked conservatism of their middle age.

Towards the end of October, on 25 or 26 of the month, Coleridge arrived by chaise at Sockburn with Cottle to see the
Wordsworths, and for the first time met the Hutchinson family,
though he had often heard the Wordsworths talk of Mary affectionately. It was another month before her sister Sara was to
become the cardinal passion in his own life. At first there was
hardly time to know the Hutchinsons, for on 27 October William
carried off his two West-Country friends on a walking tour of
the Lakes, joined by John Wordsworth who had come north to
attend Uncle Crackanthorpe's funeral. Dorothy stayed behind
at Sockburn with Mary, but received letters from the walkers,
written at Keswick on 8 November. Coleridge, excited by a new
world of beauty, wished to share it: 'Why were you not with us
Dorothy? Why were not you Mary with us?' Touchingly for
Dorothy, he appreciated the shy and silent John, his intellect,
his deep feelings, his instinct for truth and beauty: 'Your Br.
John is one of you. . . . He interests me much.'[52] William wrote
of staying at Robert Newton's inn at Grasmere. John had offered
to buy land for £40, and he had thought of 'building a house
there by the Lake side'. Dorothy's legacy would provide furniture. He added, almost casually, 'There is a small house at
Grasmere empty which perhaps we may take and purchase furniture, but of this we will speak. But I shall write again when I
know more on this subject.'[53] By 26 November, when he
returned to her, he had begun negotiations for a home at
Grasmere.

8. 'Wild, sequestered valley'

William had shown decision and energy. Within a month 'his Father's children had once again a home together', as Dorothy later wrote. Winter sunset glowed over Grasmere vale on the bitterly cold evening of 20 December 1799 between half-past four and five, when the Wordsworths arrived at their new home. It was 'a little white cottage gleaming from the midst of trees, with a vast and seemingly never-ending series of ascents rising above it to the height of more than three thousand feet'.[1] The road from Ambleside to Keswick then ran before the door, and they had already passed it without a thought on their first walk to Windy Brow, five years earlier. A few neighbouring cottages at right angles to the road formed the hamlet of Town End, the southern limit of Grasmere township, their front windows then looking directly over the lake and its reflecting mountains. Behind the house, 'our little domestic slip of mountain', with a few fruit trees, rose steeply.

Matters had moved fast since William first saw the empty cottage in November on his tour with Coleridge. On 17 December, Dorothy and he set out from Sockburn for the eighty-mile journey across the Pennines. Coach and post-chaise were far too expensive; instead Dorothy rode behind George Hutchinson on his horse and William on Sara's borrowed mare as far as Richmond. The rest of the way they walked on frozen roads, through 'driving snow-showers', at one stage covering ten miles of the high mountain road in two hours and a quarter—'by the watch . . . a marvellous feat', wrote William, 'of which D will long tell'.[2] By 20 December they were in Kendal after 'a terrible up and down road . . . buying and ordering furniture', and the same day William formally rented the cottage from its owner, Benson of Tail End.[3] It was then known simply as Town End, Grasmere, though it had once been the inn of the hamlet, the Dove and Olive Branch, from which it takes its modern name. The view from the post-chaise window was Dorothy's first sight of the home which would be, above all others, hers: a strongly-built cottage of local stone, whitewashed, with latticed windows and sturdy, thick-set chimneys.

First impressions indoors, as the darkness thickened, might have discouraged many women. 'We found no preparations except beds, without curtains, in the rooms upstairs, and a dying spark in the grate of the gloomy parlour,' wrote Dorothy,[4] but her buoyant spirits quickly rose; 'we were young and healthy and had attained our object long desired, we had returned to our native mountains, there to live'. The house would be their own, as the grander Racedown and Alfoxton could never have been. It did not take long to explore a wainscoted kitchen, once the bar of the inn, into which the front door opened, a stone-flagged room leading out of it which with a camp bed became Dorothy's bedroom, a scullery, and upstairs over the kitchen 'the Sitting Room 5 yards by 4'. In William's bedroom, the fire 'smoked like a furnace'. There was also an 'outjutting' with 'a sort of lumber room and a small, low, unceiled room'.[5] It was in no way a picturesque holiday home, but a traditional, working cottage, 'not an advertisement cottage'.[6]

The doubtful comfort of a flicker of flame in the parlour fire had been provided by the sixty-year-old Grasmere woman who stood before them. Molly Fisher, who lived with her married brother in a nearby cottage, became a legend for the Words-worths' more sophisticated friends. To Coleridge she was 'that drollery belonging to the cottage'; to Hazlitt, she was the woman who had never heard of the French Revolution. Visiting intellectuals were surprised when this rough peasant stopped them muddying the floors with their dirty boots. Like many simple people, she fell in love with Dorothy at first sight, as she stood in the cold cottage, shivering, in straw bonnet and striped gown. Invincibly ignorant, she was not a servant of the calibre of Racedown Peggy; but she was warm-hearted and asked small wages. For two shillings a week and dinner she worked two hours a day.

The rest of the work fell on Dorothy, who, in spite of a heavy cold and toothache, began with three weeks of solid hard labour, 'so much work . . . that she is absolutely buried in it', and 'so much engaged, that she has scarcely been out since our arrival', as William wrote.[7] Thus she spent Christmas Day 1799, her twenty-eighth birthday, and the following few weeks. However, before January was out she had a helpful and welcome companion in her tasks, one full of practical knowledge and resource. Her brother John had been in England since August, and was now waiting for his chance to command the East Indiaman *The Earl of Abergavenny*. He came to the cottage home for a stay that lasted eight months, until his ship sailed, early in the following year, himself its newly promoted captain. Shy as always, and half-unwilling to intrude on his admired brother's and sister's new-found happiness, 'twice did he approach the door and lay his hand upon the latch, and stop, and turn away without the courage to enter. . . . This', added Dorothy, 'will give you a notion of the depth of his affections, and the delicacy of his feelings.'[8] In the months that followed he busied himself for them with 'little schemes' ranging from extensive carpentry jobs to catching lake-trout and large pike to supplement his sister's meagre larder.[9] With his sailor's eye, used to observing every detail about the smartness and turn-out of the ships in which he served, he helped Dorothy to furnish the house and lay out the garden.

As always, the first visitor from outside the Wordsworth family was Mary Hutchinson, who appeared before the end of February for a six-weeks' stay. Scarcely had she left, on 4 April, than they had another visitor, Coleridge, who stayed nearly a month, from 6 April to 4 May. There was another short-term guest, a friend of Basil Montagu, 'so', wrote Dorothy, in one of her classic understatements, 'our cottage is quite full'.[10] Coleridge introduced his own special problems. Partly, he wished to honour his resolve, while at Alfoxton, to live always near the Wordsworths, or at any rate to have them 'within my Tether'.[11] To this end, he returned to Dove Cottage in June, and in July moved himself and his whole family to Greta Hall, Keswick, fifteen miles away, having finally decided that William 'will never quit the North of England'.[12] By now there was an additional reason to come north himself. On a return visit to Sockburn the previous November, he had fallen deeply in love with Mary Hutchinson's small, plain but lively sister, Sara: a hopeless, obsessive passion, which was, unfortunately, to colour his whole future relationship with William and Dorothy. 'A heart-rending letter from Coleridge—we were as sad as we could be. William wrote to him.' This disquietude, and William's own psychosomatic reactions to the efforts of poetic composition— 'when he uses any effort of mind—then he feels a pain in his left side'[13]—meant that the new, intimate life at Grasmere never quite reproduced the old, carefree delights of Alfoxton.

Yet these threatenings of stress had a rich by-product: on 14 May 1800 Dorothy took up her journal again. On that day, William and John had left her, to pay a visit to the Hutchinsons in Yorkshire. Lonely without her brothers, since 'poor old Molly did but ill supply to me the place of our good and dear Peggy',[14] anxious for William and, as it emerges, for Coleridge, 'I resolved to write a journal of the time till W. and J. return, and I set about keeping my resolve, because I will not quarrel with myself, and because I shall give William pleasure by it when he comes home again'. As the curious phrase 'because I will not quarrel with myself' suggests, the journal was partly to provide relief from the tensions to which Dorothy was subject. Far from giving it up three weeks later, when her brothers returned to Grasmere, she continued to keep it for two-and-a-half more years; its pages, meant only for private reading, are the prime example in our literature of a purely unconscious masterpiece.

As with her previous Alfoxton journal, it is practically impossible to put one's finger on the particular quality of Dorothy's unique gift in her Grasmere journal; the nearest analogy is the diary of Francis Kilvert, seventy years later, but even there a tinge of Victorian self-consciousness clouds the lucid picture. Though the Grasmere entries are more extensive and developed than the Alfoxton pages, which are often practically in note-form,[15] there is still not the least attempt at 'fine writing'. They mirror the places and events of the vale without imposing any of the constraints of the writer's own personality. It is interesting, now that Coleridge was experiencing the same landscape as Dorothy, to see the two describing identical places and moments, and from the same motive, or rather lack of motive: for Coleridge's Notebooks were no more intended for publication than Dorothy's journal. Though Coleridge was in many ways just as sensitive an observer, there is, without in any way decrying him, an intrusion of his personality between us and the subject observed. A favourite walk with the 'Grasmerians' was up the neighbouring Easedale Vale toward the huge multiple cataract called variously Sour Milk Force or Churn Milk

Force. Coleridge writes on 31 December 1803 of

Sour Milk Force, Langdale Pikes, Elterwater Quarries & Coniston Fells—the foot path so even on the steep breast of the Mountain, with such a precipice beneath & the tumultuous Brook at the bottom / but as you turn round & come out upon the vale, O my God! the whole white vale, from Steel Fell this way, from the Force on Easedale the River with the Mountain Islanding the half almost of the vale . . .[16]

Almost exactly two years earlier, Dorothy had written of the same experience, while walking with Mary Hutchinson:

. . . when we got into Easedale we saw Churn Milk force like a broad stream of snow. At the little foot-bridge we stopped to look at the company of rivers which came hurrying down the vale this way and that; it was a valley of streams and islands, with that great waterfall at the head and lesser falls in different parts of the mountains coming down to these rivers. We could hear the sound of those lesser falls but we could not *see* them. We walked backwards and forwards till all distant objects except the white shape of the waterfall, and the lines of the mountains were gone.[17]

Coleridge's self-obtruding exclamations—'O my God!', and so on—suffer perhaps unfairly in quotation; but there is no doubt that Dorothy's picture remains in the mind as the more transparent record.

As at Alfoxton, she seems to unite all senses, including both sight and sound. This acute awareness of sound among the vast mountain spaces pervades her most memorable moments. Sometimes, it is just used to convey a simple fact. On 15 May 1800,

I was much amused with the business of a pair of stone chats; their restless voices as they skimmed along the water following each other, their shadows under them, and their returning back to the stones on the shore, chirping with the same unwearied voice.

Sometimes, when her brothers were away, and she was alone, the lack of sound was almost too much. On 19 May

The quietness and still seclusion of the valley affected me even to producing the deepest melancholy. I forced myself from it.

At another time, her acute ear and gift for catching an unusual sound produces a little comic drama. On 1 June

I was not startled but re-called from my reverie by a noise as of a child paddling without shoes. I looked up and saw a lamb close to me. It approached nearer and nearer as if to examine me and stood a long time.

When she is expecting William's return, she hears every night-time sound in the vale. Reaching home herself on 6 June, in the evening,

No William! I slackened my pace as I came near home fearing to hear he was not come. I listened till after one o'clock to every barking dog, cock fighting and other sports.

Finally, later in the summer, after a lake expedition with her brother on a fine day, she writes:

The lake was now most still, and reflected the beautiful yellow and blue and purple and grey colours of the sky. We heard a strange sound in the Bainriggs wood, as we were floating on the water; it *seemed* in the wood, but it must have been above it, for presently we saw a raven very high above us. It called out, and the dome of the sky seemed to echo the sound. It called again and again as it flew onwards, and the mountains gave back the sound, seeming as if from their centre; a musical bell-like answering.

Such unforgettable impressions are longer and more fully developed than those of the Alfoxton journals. They also convey a whole attitude of mind and way of life. Dorothy adopts a philosophic tone in these new journals, which chimes remarkably with one particular source, and even seems to echo certain details.

This apparent source is the romantic idyll of Bernardin de St. Pierre, *Paul and Virginia*. It was evidently a part of William's favourite reading; for when taxed by Hazlitt with borrowing ideas from it, William, though denying the borrowing, showed, in doing so, a clear knowledge of the novel. He had actually written a sonnet to its most popular English translator, the 'democrat' writer Helen Maria Williams. Dorothy herself does not mention reading it now, but she lived and wrote her journal in its atmosphere, and refers to re-reading it in the later journals she kept in her old age.[18]

The novel is given an idyllic setting in a 'wild, sequestered valley' on the island of Mauritius, where the chief characters live in small cottages in a mountain vale. Two women, one an unmarried mother, the other a widow, bring up the small children, the Paul and Virginia of the title. As the children grow up to enjoy 'the pleasures of love and the blessings of equality . . . the names they learnt to give each other were those of brother and sister'. They share the work and cultivate their cottage enclosure, and 'wherever they were, in the house, in the fields, in the woods, they raised towards heaven their innocent hands'. They give names to their favourite places and relate them to human feelings, so that 'it seems to me as if a human voice issued from the stone and making itself heard through the lapse of ages addressed man'. They advocate simple life and dislike the idea of sophisticated society. They dwell on the perfections of fraternal love. Virginia says, 'Ah! if it was again possible to give me a brother, should I make choice of any other than you?' and Paul exclaims, 'She is everything to me, riches, birth, family, my sole good!' Adult sexuality does not threaten this ideal relationship; indeed it is ruthlessly avoided by the drowning of Virginia in a shipwreck and the consequent death from grief of Paul.

There are exact parallels between this textbook of the Romantic movement, and the ideas, incidents, and expressions of life at Grasmere. Indeed it recalls Dorothy's whole life, in which, she wrote later, 'fraternal affection . . . has been the building up of my being, the light of my path!'[19] Perhaps among the most striking of the parallels is the Wordsworths' habit of naming natural features after themselves and their relatives and close friends. Just as with Paul and Virginia, rocks are inscribed with the initials of the Wordsworth circle. In the first summer of their Grasmere enthusiasm, a large rock on the banks of Thirlmere, on their way north over to Keswick, was cut with the initials W. W., M. H., D. W., S. T. C., J. W., S. H., to commemorate William, Mary Hutchinson, Dorothy, Coleridge, John Wordsworth, and Sara Hutchinson, afterwards referred to by Dorothy as The Rock of Names or Sara's Crag; while two years later, when the letters had been re-carved and deepened by them, Dorothy recorded in her journal, 'I kissed them all'.[20] A grove of fir-trees on the way to Ambleside became known as 'John's Grove' because John Wordsworth

By pacing here, unwearied and alone,

had trodden out a path there. Nearly opposite, a gate from which Sara Hutchinson watched a favourite view became known by them all as Sara's Gate. William, in this happy summer of 1800, wrote a set of poems about these natural features called 'On the Naming of Places', from which Hazlitt assumed, probably with justice, that the Wordsworths had taken the idea from St. Pierre's book.

Nor were these resemblances confined to natural, physical features, such as rocks, trees, gates, and so on: the attitude of the book toward matters of the mind was repeated in life at Grasmere. The avowed moral of the book, 'to inspire a taste for the blessings of nature', the love of labour and the dread of riches', was faithfully followed. On 2 October 1800, Dorothy entered in her journal that she, William, and their acquaintances the Lloyds, who had also come to live near Grasmere, 'had a pleasant conversation about the manners of the rich—avarice, inordinate desires, and the effeminacy, unnaturalness, and the unworthy objects of education'.[21] As well as such rather curiously named 'pleasant conversations', there were moments of solemnity. Just as Virginia, in the book, felt 'all which surrounded them, one holy temple', so Dorothy, walking alone, found that 'Grasmere was very solemn in the last glimpse of twilight; it calls home the heart to quietness'.[22] Nature, indeed, in St. Pierre's book, gives Virginia all she wants, and she is content 'to enjoy these pure and simple pleasures'. So with Dorothy, when on the evening of 2 June 1800, the friendly postmistress of Ambleside walks with her as far as Rydal, in case she should be lonely: 'This was very kind, but God be thanked, I want not society by a moonlight lake.'[23]

Just as Paul and Virginia find 'so much simplicity and good faith' in the island's peasants, so Dorothy's democratic enthusiasm found everything to praise in 'our neighbours of the lower classes'. She reported to Jane Marshall, 'they are excellent people, friendly in performing all offices of kindness and humanity and attentive to us without servility'.[24] Coleridge had the impression that Wordsworth 'and his sister are exceedingly beloved, enthusiastically' by the local people.[25] There is evidence, though, that they seemed odd. To struggling farmers and labourers they were rich, since they did not need to work. Yet they tramped for miles, at all hours, by day or night, 'the

moon just setting as we reached home', [26] or in driving rain 'to
observe the torrents'. [27] Wordsworth 'went into the wood to
compose' out loud and, Dorothy said, might stand for half an
hour under an umbrella in road or field contemplating the sodden landscape. [28] Their clothes and shoes were tattered with
rough usage; William in any case wore his brother Richard's
cast-off coats. They got up and went to bed when they felt like
it. On a fine night Dorothy 'lingered with Coleridge in the
garden. John and Wm. were both gone to bed and all the lights
out.' [29] When Coleridge arrived late, they 'sat and chatted until
half-past three in the morning, Wm. in his dressing gown'.
Over-strained by composition, Wordsworth 'rose late', 'lay
long in bed', or sometimes 'made himself very ill', and took to
bed in the daylight where Dorothy sometimes 'read to him to
endeavour to make him sleep'. Visitors came and went at the
cottage without ceremony. Mary, and later Sara, Hutchinson
arrived as extra women in the household. Dorothy was 'roused
by a shout' that Anthony Harrison of Penrith had called in passing, or old Mr Sympson 'called in for a glass of rum just before
we went to bed'. [30] Coleridge arrived like one of the family,
'when we were at dinner very wet', or Dorothy 'broiled Coleridge a mutton chop which he ate in bed'. Hospitable and un-
selfconscious as a child, she did not think of village gossip, that
powerful force in country life. Yet there is evidence that this
free and easy household aroused suspicion, however unjusti-
fied, in Grasmere, just as it had at Alfoxton, and of an even
more sinister nature.

The nature of the gossip was revealed on one particular
occasion, during the lifetime of Dorothy, by Thomas De
Quincey. De Quincey is sometimes a dubious source for matters of fact, but in this instance there are circumstances that
lead one to believe him. A young disciple of Wordsworth, he
had at last achieved his ambition of becoming acquainted with
the poet at Dove Cottage. Moreover, when the Wordsworths
quitted the cottage, he took over their tenancy. A few years
later, he became the lover of a Grasmere village girl, Peggy
Simpson, whom he subsequently married. He was therefore
well placed to know what, for many years, had been the local
gossip about the Wordsworths and their household.

In 1821, just over twenty years later, De Quincey was contributing his famous *Confessions of an English Opium Eater* to the *London Magazine*. He was frequently in London, and attended the social evenings of the brilliant set of writers gathered by the publishers of that magazine. After one of these, on 6 December, he went on to the chambers in Hare Court, Inner Temple, of the publishers' legal and literary adviser, Richard Woodhouse. The 'little lawyer', so-called from his small stature, had a gift for extracting confidences, aided by the excellent claret he kept. Two years earlier, under very much the same circumstances, he had won the confidence of John Keats so fully that the latter had talked for six hours or more. De Quincey was no less susceptible to Woodhouse's friendly charm, and began to unburden himself specially about slanderous gossip in the Lake District. As Woodhouse noted down,

Speaking of the characters of minds of different people, & indeed of various whole classes, he took notice that he considered the minds of the people in his own neighbourhood as being particularly gross & uncharitable—That they were fond of retailing anecdotes however horrible as true without even taking the trouble to ascertain their foundation, or caring at all whether they were true or not. . . . the opium eater mentioned several stories, entirely groundless, & carrying in their very horror an appearance of their falshood & absurdity—

So much for Dorothy's optimistic and innocent description of her Grasmere neighbours as 'excellent people, friendly in performing all affairs of kindness and humanity'. What they were actually saying about the Wordsworths was, as he went on, revealed by De Quincey.

First, there was a rumour that one of Wordsworth's children was not really his, but was the offspring of De Quincey himself and Mrs Wordsworth. Having disposed of this piece of total improbability, De Quincey proceeded to relate an even more damaging, sensational, and apparently widely held piece of Grasmere gossip.

Again there was an unnatural tale current, & which the O. E. had heard even in London, of Wordsworth having been intimate with his own sister—The reason for this story having birth seemed to be that Wordsworth was very much in the habit of taking long rambles among the mountains, & romantic scenes near his habitation—his sister, who is also a great walker used very frequently to accompany him and indeed does so still—It is Wordsworth's custom whenever he meets or parts with any of the female part of his own relations to kiss them—This he has frequently done when he has met his sister on her rambles or parted from her and that in roads or on mountains, or elsewhere, without heeding whether he was observed or not: and he has been perhaps seen by hinds and clowns or other persons who have repeated what they have seen:—and this simple fact, occurring probably under the eyes of those (and O. E. says he has met many such, even in the upper and better-informed classes) who have not the slightest idea of pure love for any one or of that fine tie which forms the affection between a brother & a sister, has been made up into the abominable accusation bruited about, to his prejudice amongst his coarse-minded neighbours.[31]

Several facts need noting about De Quincey's account. First, he of anyone, married to a local villager, should have known what village people were really saying. Second, he does not confine the rumours to 'hinds and clowns', but, more widely, to 'the upper & better-informed classes' and has heard it 'even in London'. Finally, although he himself had every reason to believe the worst about William and Dorothy, since both had treated the sexual peccadillo leading to his own marriage with a notable lack of charity, he comes down firmly on the side of their innocence, and, indeed, provides a rational explanation for the rumour. There is no reason to doubt De Quincey's sincerity, especially under the circumstances in which he conveyed this information to Woodhouse in the intimate atmosphere of the lawyer's rooms.

The story, in fact, disappeared for 150 years into Woodhouse's private and never-published papers. It was not until the second half of the twentieth century that it reappeared in a new form. The hypothetical incestuous relationship between William and Dorothy, far from being a modern idea, was something held, in Dorothy's own time, as an unfounded piece of gossip by those whom De Quincey castigates as 'particularly gross & uncharitable'. Nor should De Quincey's own explanation of William's kissing-habits be dismissed as too simplistic by a more sophisticated age. Wordsworth *was* in the habit of kissing women, even outside his own family, as a normal gesture of parting.

In fact, although the Wordsworths may have seemed odd to the neighbourhood, as visiting intellectuals in the country often

do, there was a great deal of normality in their lives. For one
thing, this was ensured by the sheer amount of work to be got
through. All through June, July and August 1800, Dorothy
had to copy and recopy the poems of William and Coleridge,
including a much-revised *Ancient Mariner* for the latter, and, in
September for William, a whole new long Preface for the fresh
second edition, which Longmans were now bringing out, offering William £80 for the new publication.[32] The time she shared
with her two brothers was full of hard effort. Fishing expeditions were a necessity as well as a pleasant way of spending this
exceptionally fine summer. Dorothy's happy settling-in at
Dove Cottage was, indeed, largely due to John's presence, and
she felt a deep loss when he departed in autumn 1800. William
and Dorothy said farewell to him at Grisedale Tarn, on a
shoulder of Helvellyn, and saw him descend into the next
valley.[33]

We were in view of the head of Ulswater, and stood till we could see
him no longer, watching him as he *hurried* down the stony mountain.

John was a brave and modest sailor, who thought himself lucky
in his 'very *noble ship*', its officers and midshipmen 'the finest
lads I ever saw'. Humbly, he asked William what books he
ought to take to sea; 'I do not know what I should like myself'.[34]
The tragedy that finally overtook John coloured Dorothy's
memories of this happy, shared summer for the rest of her life.

Yet this first year at Grasmere had other associations and
friendships, not so emotionally fraught but in many ways deeply
satisfying. 'We are also upon intimate terms', wrote Dorothy
to Jane Marshall,[35] 'with one family in the middle rank of life,
a Clergyman with a very small income, his wife, son and
daughter.' This was the Reverend Joseph Sympson, the Vicar
of Wythburn on the way to Keswick, who lived at High Broadrain, a farmhouse near Dunmail Raise. A hearty octogenarian,
'he goes a fishing to the Tarns on the hill-tops with my Brothers,
and he is as active as many men of 50'. Though the Wordsworths
seldom attended his tiny church nestling under Helvellyn, he
formed the habit, with his daughter Margaret, of calling on
them for tea, and asking them back, at least once a week. Miss
Sympson, a few years older than Dorothy, was a companion
with whom she was glad to share her otherwise solitary walks,

when William and John were away; she too was friendly and solicitous in providing small luxuries from the farmhouse, such as plentiful gooseberries, 'and cream', as Dorothy exclaimed.

Late in August 1800 a remarkable pair of neighbours came to dine, walk round Rydal in the rain, and stay for the weekend. Thomas Clarkson was, like William, a member of St. John's, Cambridge. As a young graduate he had read in the Senate House, to loud applause, a Latin essay on slavery; and riding to London afterwards vowed to devote his life to ending the slave trade.* Nine years of exhausting travel broke his health, and he retired to build a house on a beautiful site at Eusemere Hill, at the east end of Ullswater. Here he brought his young wife in 1796 and their only child was born. Catherine, daughter of a wealthy yarn-maker, William Buck of Bury St. Edmunds, was a year younger than Dorothy, not formally educated but well-read and brilliantly amusing beside her serious husband. On many return visits to Eusemere, William and Dorothy would walk over the steep pass to Patterdale, at the near end of Ullswater, and the Clarksons would send a boat to ferry them the whole length of the lake. At Eusemere, Catherine Clarkson entertained the Wordsworths with stories of the charitable curate's wife who 'killed a pig with feeding it overmuch' and thanked God 'it did not die *clemmed*'.[36] Dorothy, in return, took an interest in little Tom's teeth, and the planting of shrubs in the lakeside garden. Mrs Clarkson, trusted in 1802 with family secrets, William's courtship and the existence of his little French daughter, responded with 'a very kind affecting letter'. The easy friendship of the two women, one so amusing, the other so emotional, deepened into a lifelong love.

With visits to and from fresh friends such as these, and from old ones, Williams's college-friend Jones, and Jane's husband Mr Marshall, all woven into the pattern of a ceaselessly busy life, Dorothy's journals seem more than ever miraculous. Yet these journals, for all their almost magical quality, chronicle the facts of that life: the practical business and never-ending effort of one woman's existence.

* A monument on the hill at Wades Mill, Herts, marks the spot.

9. *'Plenty of Business'*

When the *Journals of Dorothy Wordsworth* were first published in two volumes in 1897, the editor Professor William Knight, author of an 1889 *Life of William Wordsworth*, in which he had already used lengthy extracts from Dorothy, wrote apologetically of the 'numerous trivial details' they contained. Some were justifiable to illustrate approvingly the 'plain living and high thinking' of the Wordsworth household, but 'there is no need to record all the cases in which the sister wrote, "Today I mended William's shirts"', since these facts had no literary value.[1] Seen through Victorian eyes the sight of a great writer's sister doing menial work was distasteful. Later readers are more likely to see the interest of Dorothy Wordsworth's writing as social history. Many accounts exist of life in great households, but she shows the day-to-day existence of a family of small means, in a region traditionally poor and isolated, forced to be self-reliant in provisioning, baking, brewing, repairs, laundry, and gardening, not to speak of sick-nursing, birth, and death.

The romantic simple life apparently assigned the duties of housekeeping as firmly to woman as any traditional household. Dorothy Wordsworth accepted as natural the continual labour which formed the backbone of everyday existence. Indeed she clearly took pride in her skills. The housewife later lost status with the arrival of manufactured goods, imported foods, and a rising population of servants, but in 1800 she had the satisfaction of knowing she was essential to survival. The lady of leisure had not yet appeared in middle-class circles, and Dorothy was critical of Sara Coleridge's genteel pretensions: 'she is to be sure a sad fiddle faddler,' she wrote briskly to Mary Hutchinson.[2] For herself she refused to admit conflict between her duties as housekeeper to William and any social or intellectual interests. Journals or letters, for instance a letter to Coleridge on 9 December 1801, might be 'written in the morning by the kitchen fire while the mutton was roasting'. The kitchen was warm but dark, and wax candles at five shillings a pound were sparingly used in middle-class households; perhaps the blink of home-made tallow dips contributed to the marked deterioration

of Dorothy's once neat schoolgirl hand. Her own tastes were
frugal; alone at home in February 1802, although the frost was
so hard that her 'inside was sore', she chose to eat 'a little bit of
cold mutton without laying cloth and then sate over the fire
reading'.[3] Housekeeping for others, by contrast, took first
place. As she later wrote, 'when I am put off my usual course
though having plenty of leisure I am in a state of mind that
particularly unfits me for writing', unlike times 'at home with
plenty of business'.[4]

The business of housekeeping was less hard than it would
have been a generation earlier. Riding through this remote
countryside in the year of Dorothy Wordsworth's birth, Arthur
Young had found it 'totally uncultivated'. For lack of winter
feed, farmers held a great autumn slaughter, after which only
smoked or salted meat was eaten. Wheat would hardly grow on
the black uplands and housewives baked bread from oats or
barley harvested with a sickle and thrashed with a flail. By the
time the Wordsworths moved to Grasmere, the agricultural
revolution had introduced new fodder: clover, beans, and roots
to feed stock in winter; milk, butter, and meat were now available all the year round for those who could afford to buy them.
Yet, as her journals show, Dorothy Wordsworth had to work
steadily and hard to feed a household with frequent visitors on
a small income. It was a romantic ideal to live simply, upon the
bounty of nature, but for her it was also an economic necessity.

William kept up essential fishing, and she went with him,
though sitting still by the lake, 'my head ached with cold', and
she confessed she still thought of John 'whenever we see the
shape of a trout'.[5] They had already borrowed a boat from one
neighbour, pike-floats from another, and rowed on the lake
together; even in June it was 'extremely cold . . . the lake clear
to the bottom but saw no fish'. On another cold morning they
set pike-floats, and next day 'returned to dinner, 2 pikes boiled
and roasted'.[6] Dorothy stuffed pikes to put in the oven, or 'put
up the bread with a few baked trouts'. Sometimes whole parties
'went upon the water, after tea' with the boat owner, or
'amongst us caught 13 Bass'. When the Coleridge family were
expected for a visit William and John made a special fishing
expedition to Rydal. With salmon reaching one shilling a
pound in Kendal market one can see why.

Another source of free food was the garden, where from the first day they cultivated vegetables. On the second day of her Grasmere journal, Dorothy 'Hoed the first row of peas, weeded, etc., etc.'. Old Molly Fisher weeded turnips while her brother John stuck peas. By early June Dorothy sowed French beans, planted broccoli, and watered on dry evenings.[7] Soon the cottage was 'covered all over with green leaves and scarlet flowers, for we have trained scarlet beans upon threads, which are not only exceedingly beautiful, but very useful as their produce is immense'.[8] By August, there were not only 'peas for dinner', but baskets of peas and beans to take as gifts to old Mrs Sympson at Broadrain. The glut of gooseberries drove Dorothy on 7 August to boil three panfuls 'all good measure', with sugar to preserve them; evidently she was proud of the recipe, since she entered it in her journal.[9] In September, she still 'walked with French Beans to Mr. Simpson's'.

Presents of food were common in all social classes. John sent a box of tea,[10] and Mr Griffith a barrel of best flour, which cost 2*s*. 8*d*. to 3*s*. 2*d*. a stone at Kendal. Produce could not easily be preserved, so it was shared among friends and neighbours when plentiful. Friends were generous to Dorothy. Mr Clarkson, who knew the state of the Wordsworth finances, gave her a Michaelmas goose in 1801. Dorothy at once thought of her poorer neighbour who was ill, and 'sent Peggy Ashburner some goose. She sent me some honey with a thousand thanks.'[11] In January 1802 Mr Clarkson provided a turkey, of which the gizzard made a whole meal for two and the giblets a pie. He even appeared on one visit with 'a calf's head in a basket'. Mrs Sympson sent 'pork to dinner', Janet Dockeray 'a present of Eggs and Milk', and Mrs Oliff, the farmer's wife, 'some yeast', all returned in kind. The tradition of mutual help, so democratic and so near to nature, fulfilled the romantic ideal of country life. Just so had the families of Paul and Virginia lived, without luxuries which 'their industry could not supply', cultivating the land with their possessions 'in common'.

The work of baking continued all the year round; home baking was a great economy and most North-Country kitchens contained a kneading-trough for bread, as well as a large pot for porridge. On baking day the oven was well filled: 'made tarts, pies etc.', 'baking bread, pies and dinner', 'bread in the

morning', 'baking bread apple pies and giblet·pie—a bad
giblet pie'.[12] The oven, affected by wind, was not foolproof.
William recorded gingerbread burnt black as coal by too much
heat on Molly's advice, and also a 'very bad dinner' which
Dorothy 'had the vexation' of giving to one of his friends, and
which he remembered ten years later.[13] Other guests were
happier. When Sara Hutchinson came in December 1800, the
two women enjoyed 'a grand bread and cake baking' with 'little
loaves', and next day 'Supped upon hare'.

Even with home produce, gifts, and continuous baking there
were occasional shortages. In March 1802, when William was
in Keswick, Dorothy 'got my dinner two boiled eggs'. In May
the same year William and Coleridge had 'mutton chops and
potatoes', with no other vegetables, for supper. In June Dorothy
and William 'ate some Broth for our suppers' and ten days
later 'got no dinner, but gooseberry pie to our tea'. In winter
they finished off whatever was left over. 'I am going to take
tapioca for my supper, and Mary an egg, William some cold
mutton.'[14] Dorothy often ate 'hasty pudding', a staple food of
the poor in the North of England. This was a gruel of oatmeal
boiled in salted water, which would feed one person for less
than a penny. Nothing could show more clearly the determined
simplicity of her tastes.

The sole domestic helper in the early Grasmere years remained 'old Molly'. By the standards of the time and place
sixty really was old; she was 'very ignorant, very foolish and
very difficult to teach'. Dorothy almost despaired, but was
determined to persevere because 'she is much attached to us
and honest and good as ever was a human being'.[15] At first she
could only light fires, heat water from the syke to wash dishes,
and do the regular Saturday 'scouring', after which Dorothy
gave her a meal. Gradually she learnt to boil vegetables and
'watch the meat'. After eighteen months, Dorothy told Mary
she 'manages the oven entirely and as well as I can'. Left alone
she 'shook the carpet and cleaned everything upstairs'. She
gloried—her own word—in the household possessions, and
would have ventured her own life savings, more than seven
pounds, in a trading voyage with her favourite 'Maister John'.
Tears came into her old eyes when she talked of him, far away
at sea. After a life of poverty and toil she was happy to be

important to the Wordsworth family. 'Aye Mistress,' she said, 'them 'at's low laid would have been a proud creature could they but have [seen] where I is now fra what they thought mud be my doom.'[16] Dorothy was touched to see Molly so happy in her work. Belief in simple people was not a theory to her, but a living reality.

Dorothy Wordsworth was a resourceful and practical odd-job woman, unafraid of work with her hands; indeed labour formed part of her romantic ideal of country life. Early days at Grasmere 'in such confusion as not to have had a moment's leisure', gave scope for these inherited Yorkshire talents, and the journal suggests the range of her tasks. 'We put the new window in',[17] 'On my return papered Wm's room',[18] carried coals with 'a habit of attention and frugality'; or while mending old clothes 'Bound carpets'. She bottled rum, perhaps brought by sailor John, since there is no record of buying it. Hemming curtains, sheets, towels, and table-cloths by hand was a matter of course, also mending and 'making my shifts'. On the very fine evening of 5 August 1800, she 'Sate on the wall making my shifts till I could see no longer'. Ten days earlier she had 'made my shoes', and sat upon the same wall 'till near dark' finishing them. In August she bought sacking and in September was 'busy all the morning making a mattrass. . . . Finished the mattrass, ironed the white bed in the afternoon' and put it up.[19] 'Nailed up the beds', 'fixing up vallances', and 'sweeping chimneys' were routine duties. She laid matting on the stone floors of the room next to the kitchen, which became her bed-room. 'The bed though only a camp bed is large enough for two people to sleep in', she told Jane. Upstairs was the sitting-room, with a view over the lake, William's bedroom with two single beds, 'and a small low unceiled room which I have papered with newspapers'[20] as a home-made insulation against cold. In this unheated small room, above the icy larder over the syke, she later slept herself. When work at home was finished, she was called upon in family removals or spring-cleanings, yet still found time to make a seat beside the path through the woods to Windy Brow at Keswick.

Dorothy's eager dedication withstood even the hazards of laundry in the Lake Country. She told Jane in September 1800 that 'We have never hired any helpers either for washing or

ironing'. Old Molly washed clothes once a week 'apart from
the house' which 'makes it very comfortable', and Dorothy
helped 'at the great washes about once in 5 weeks'.[21] Reality,
as the journals show, was sometimes less comfortable than she
would admit. In the South, the work had been straightforward,
if hard: 'one day to wash, on the next we have got the clothes
dried and on the third have finished ironing'. At Grasmere the
weather was an unpredictable partner, sometimes 'a boisterous
drying day' at others 'Dullish, damp and cloudy—a day that
promises not to dry our clothes'.[22] 'Went often to spread the
linen which was bleaching—a rainy day and a very wet night,'
she wrote, or 'We dried the linen. Frequent threatening of
showers.'[23] Even summer could produce 'A very rainy morning . . . Molly began to prepare the linen for putting out, but it
rained worse than ever,' Dorothy wrote on 2 July 1802. Sometimes washing stayed for three or four days. Water had to be
carried from the syke, 'that diminutive beck where we get our
water', and heated over a fire, and the work, done by hand,
was slow. 'I was starching small linen all the morning . . . it
snowed,' she noted on 15 February 1802. On other days 'I
ironed till dinner time' or 'I was ironing all the day till tea-
time'; 'ironed the white bed in the afternoon' or simply 'ironed
all day'. It was an achievement to be 'Putting linen by and
mending', when the wash was done.

Dorothy Wordsworth was a flower gardener, of consistent
and discriminating taste. Her ideal garden borrowed from
nature and created harmony between house and countryside.
At Alfoxton she had made homely cottage gardens 'the object
of our walk', while later in Scotland she lamented the 'unnaturalness' of a modern garden with flower borders, and the whole
'miserable conception of *adorning*' the ruined castle at Bothwell.[24]
At Town End for the first time in her life she and William
could create their own ideal garden. First they enclosed the two
or three yards between their door and the road with upright
stones 'to make it more our own', and pulled down a fence
which divided the cottage from the steep orchard behind. Then
Dorothy set out, like the hero of *Paul and Virginia*, to transplant
her flowers from the wild. 'I carried a basket for mosses', she
wrote in mid-May 1800, 'and gathered some wild plants. Oh!
that we had a book of botany.'[25] A week later she 'Brought

down Batchelors' Buttons (Rock Ranunculus) and other plants'. John joined in these forays. 'We went up the hill to gather sods and plants and went down to the lakeside and took up orchises.' He turfed the well, where she planted London Pride, and 'many things on the Borders'.[26] Returning late with lemon thyme, too eager to wait, she planted by moonlight. Other triumphant transplantings were wild columbine, fox-gloves, and wild snowdrops. Gifts from country gardens were welcome; visiting Jenny Dockeray's farmhouse on the Easedale road, Dorothy 'went into her garden and got white and yellow lilies, periwinkle, etc.', which she successfully watered in.[27] The Sympsons contributed sunflowers, mulleins, and pansies.

Dorothy planted not to make a show of flowers but to share in the work of nature, rooting plants where they could live out their lives, roses with the scarlet beans against the cottage walls and honeysuckle twining itself round the yew tree.[28] On a sum-mer day of incessant rain, she felt content, 'the rain falls so gently and the air is so mild and all living things seem so much to rejoice in it that one can hardly find in one's heart to wish it over'.[29] She had her share of gardener's vanity, describing to Jane Marshall the 'small orchard and smaller *garden* which as it is the work of our hands we regard with pride and partiality'. True, the orchard was very small, but delightful from the 'excessive beauty of the prospect' over lake and fell. On 9 June 1800 Dorothy was sitting on the turfed wall when a coroneted landau went by; she recorded with satisfaction that 'the ladies (evidently Tourists) turned an eye of interest upon our little garden and cottage'.[30] To such people of taste, Dorothy's neat garden, she knew, announced her as respectable, though her beans and honeysuckles, humble person's plants, showed her as poor as her neighbours. In her flower garden, as in much else, Dorothy exemplified the fashion of the moment.[31]

There is no sign in the Grasmere journals of any distinction between work and leisure; housekeeping, gardening, and walk-ing wove their way through the changing seasons. Dorothy Wordsworth's reading, which she loved, was casual. At first she was too busy to read the German books Coleridge lent, though in March 1802 she read 'German especially Lessing' to translate. When William and John were away in summer 1800, she read entire Shakespeare plays in the evenings; later in 1801 she and William read Chaucer aloud to each other for him to translate. In bed, after taking laudanum for toothache, she indulged herself in *Tom Jones*. She read aloud to William the eleventh book of *Paradise Lost* in February 1802 and both 'melted into tears'; on Christmas Eve by the fire they read '*L'Allegro*' and '*Il Penseroso*'. She read William to sleep with Spenser, and Spenser's '*Prothalamion*' for herself, walking back and forth in the orchard on an April morning.

The historical interest of Dorothy's Grasmere journal is not purely domestic. She was a faithful witness to the lot of the poor at a time of exceptional hardship, 'exposed to calamities, which no human foresight can avert'. The Wordsworths never refused hospitality or money to the wayfarers: unemployed journeymen and labourers, discharged soldiers and sailors, pedlars, gypsies, and beggars, who constantly passed their door. The times conspired against these ragged walkers. Domestic spinners and weavers found their hand work undercut by machinery, which could be minded by eight-year-olds at two shillings and sixpence a week.[32] Farm labourers lost cottages, vegetable plots, and rights to grazing or fuel-cutting under farm enclosure schemes. Ten thousand acres in Westmorland were enclosed between 1760 and 1800 and gangs of 'wallers' camped out dividing the open fells with a mesh of dry-stone walls. This was the subject of complaint by a neighbour, old Molly's brother John Fisher. On 18 May 1800, he overtook Dorothy on the road to Ambleside, 'talked much about the alteration in the times and observed that in a short time there would be only two ranks of people, the very rich and the very poor, "for those who have small estates" says he "are forced to sell and all the land goes into one hand"'.[33] Wartime taxes drove simple necessities like soap or candles beyond the means of poor housewives. Most damaging, perhaps, was the naval blockade and the soaring price of corn, which can be seen in the rates of poor relief, commonly calculated on the price of bread. In January 1800 the 6*d* loaf stood at 17*d*. 'I sincerely wish', wrote Parson Woodforde on 31 January, 'that it might be cheaper e'er long for the benefit of the Poor who are distressed on that account.' Necessity drove many on to the roads in search of a living; a Swiss traveller, Barretti, found in England 'beggars enough to fill a province'. These were the travellers described by Dorothy Wordsworth, not as a social problem, but as individuals, for whom she felt personal sympathy and respect.

They were a varied company, even in their troubles. In May 1800 a young woman from Manchester begged at the door. 'She had buried her husband and three children within a year and a half—all in one grave—burying very dear.'[34] Four days later came a little girl who had slept out—'her step-mother had turned her out of doors'. On 27 May 'a very tall woman' in long brown cloak and white cap called at the door with a bare-footed child. 'I gave her a piece of bread.' Afterwards she saw the husband feeding his donkeys by the roadside and, unforgettably two boys 'at play chasing a butterfly. They were wild figures . . . without shoes and stockings; the hat of the elder was wreathed round with yellow flowers, the younger whose hat was only a rimless crown, had stuck it round with laurel leaves.' They begged from her with professional skill.[35] The family is recalled in Wordsworth's poem 'Beggars'. Another subject for poetry was 'an old man almost double', encountered in the dusk on 3 October 1800. 'His trade was to gather leeches, but now leeches are scarce and he had not strength for it. He lived by begging.' Nearly two years later, this man was to become the semi-symbolic figure of William's *Resolution and Independence*.

Not all the travellers lived by begging. A sturdy woman from Cockermouth had travelled over the mountain for thirty years selling 'thread, mustard, hardware, etc. . . . she does not mind the storms if she can keep her goods dry.' Some passed swiftly, like the 'merry African from Longtown'. Others lingered to talk of their downfall like the woman on a frosty 27 November 1800.[36] ' "Aye" says she, "I was once an officer's wife, I, as you see me now. My first husband married me at Appleby; I had 18 £ a year for teaching a school, and because I had no fortune his father turned him out of doors." ' She described how he had been shot, and how she walked the roads with '*that man*', her lame second husband. A few days later, Dorothy and Mary, returning from the post, were overtaken by two drunken soldiers; far from being frightened, she found them 'very merry and very civil . . . they were honest looking fellows'.[37]

For brother John's sake, Dorothy had a special fondness for sailors. Meeting a seventy-five-year-old sailor before Christmas on White Moss, she let him pass. 'He said nothing, and my heart smote me. I turned back and said, "You are begging?" "Aye" says he. I gave him a halfpenny.' Questioned by William, the old man said, 'I have been 57 years at sea, 12 of them on board a man-of-war under Sir Hugh Palmer.' 'Why have you not a pension?' 'I have no pension, but I could have got into Greenwich hospital, but all my officers are dead.' She described his fresh cheeks, cast-off gentleman's clothing, and bow legs in loving detail.[38] The sailor was a tidier figure than one seen 'sitting in the open field upon his sack of rags the old Ragman that I know. His coat is of scarlet in a thousand patches. His breeches' knees were untied.' He asked, in dialect, 'Is there a brigg yonder that'll carry me ow'r t'watter?'[39] On 15 March 1802 a young sailor sat for two hours by the kitchen fire, telling stories of his life at sea. 'He had been on board a slave ship . . . where one man had been killed, a boy put to lodge with the pigs and half eaten, one boy set to watch in the hot sun till he dropped down dead. . . . He said he would rather be in hell than be pressed.' Dorothy found 'He was excessively like my brother John'.[40]

Wayfarers with children found a specially kind welcome. 'The dismal sound of a crying infant' drew her to the window. One woman often called, 'and her little boy—a pretty little fellow, and whom I have loved for the sake of Basil—looks thin and pale. . . . The child wears a ragged drab coat and a fur cap, poor little fellow, I think he seems scarcely at all grown since the first time I saw him'—the stunted growth of near-starvation. Destitute herself from childhood, and at thirty still dependent on the charity of friends, she wrote, 'I could not help thinking that we are not half thankful enough that we are placed in that condition of life in which we are.'[41] It seems that she never quite lost the traditional belief in providence which she had been taught as a child.

It would be rash to speak too confidently of Dorothy Wordsworth's religious life, since for many years she chose to speak little of it herself. This reticence lasted into middle age, when conventional displays of piety still jarred upon her fastidious taste. She was uneasy at Mrs Klopstock's letters in translation with 'God in every fourth line of a page',[42] and smiled at acquaintances who always spoke gravely of 'the Lord's Day' and

5. 'when I first visited the Wye, and all the world was fresh and new': Tintern Abbey

6. 'still reclusion': Grasmere

'the Sabbath' for Sunday. She was worried when Catherine Clarkson risked cold at church on four successive Sundays. When another friend of the Clarksons took 'a decided turn' towards religion, Dorothy regretted it; 'she has honesty and sensibility enough to have been good in another way'.[43]

The Alfoxton and Grasmere journals make clear that formal prayers or church services played little part in her life, although she once went twice to church in one Sunday with the clergy-daughter Miss Sympson, when a christening 'kept us very late'.[44] At Alfoxton she had spent the morning walking in the woods on Easter Sunday. Successive Sunday entries in the Grasmere journals show the Wordsworths untrammelled by church-going. In 1800 on 7 June they did not get up until ten o'clock and spent the day writing letters. On 19 October they again 'rose late and walked directly after breakfast'. On 29 November Dorothy walked with Sara Hutchinson to enjoy the snow. The following year she noted of Sunday 18 October simply 'I have forgotten'. The last Sunday in November she chose for a large baking, an unthinkable activity by the conventions of her later years. Sunday was also a favourite day for long expeditions on foot. On 31 January 1802 they 'walked round the two lakes', and sat by a stone 'close to Mary's dear name which she had cut herself', watching the breezes brush the surface of the water, and the Islands reflected in a 'rich yellow light'. Sunday, 28 March they also spent on the road, walking to Keswick to visit Coleridge, less idyllically. 'Arrived wet to skin.' Good Friday, 16 April 1802 they spent on the road from Eusemere home to Grasmere over Kirkstone Pass, among sights and sounds which inspired Dorothy to some of her most magical writing. 'I hung over the gate,' she added simply, 'and thought I could have stayed for ever.'[45] There is no suggestion, though, that she felt it a specially hallowed day. Often Sunday was simply a day for odd jobs; she sat by the fire and read or 'copied and stitched poems'. As summer came on they sat talking in the orchard. On 13 June 1802 she 'copied poems and wrote letters. W. washed his feet.'

This casual flouting of convention marked a deliberate break with Dorothy Wordsworth's upbringing and her young womanhood in Forncett Rectory. Yet the same journals carry mysterious echoes of the holiness she found in everyday life and humble people. The sights and sounds of the Lake Country stirred her nature to its depths; the daily entries, so simple and truthful in their account of common things, distil and deepen ordinary life to a spiritual experience. There is no sign Dorothy Wordsworth took a self-conscious decision like Rousseau that walls formed a barrier and she would worship by 'contemplating God in the glories of creation'. She responded to life with reverence, and abandoned the forms of worship, not the reality. She made no formal declation of belief, like Wordsworth's own

> In the primal sympathy
> Which having been must ever be;

Her moments of vision were so spontaneous, recorded so simply for herself and William alone, that they seem almost unconscious. Outwardly, at this time she shared some of his dislike of the clergy. When she went on 3 September 1800 to the funeral of a poor woman buried by the parish, she noticed that the curate for the absentee rector 'did not look as a man ought to do on such an occasion'; she had seen him half-drunk in a pothouse the day before, which was Grasmere Fair. Yet she was deeply moved when they carried the coffin out of the dark house into the sun, shining over the vale, 'so divinely beautiful as I never saw it. It seemed more sacred than I had ever seen it, and yet more allied to human life. . . . I thought she was going to a quiet spot and I could not help weeping very much.'[46] The church itself, to which they so seldom went, was hallowed; looking out in November moonlight she saw 'The Lakes beautiful. The church an image of peace.' Sacred places kept their traditional meaning for her; walking in Scotland in 1803 by the ruined priory near Hamilton she felt 'Nothing can be more beautiful than the little remnants of this holy place'.[47] At the sight of a beggar woman and child who came to the door in frost and snow, she wrote spontaneously, 'We do not so often bless god for this as we wish for this 50 £ etc., etc. . . . This woman's was but a *common* case.'[48] She accepted these harsh decrees of heaven, apparently without question.

The sense of holiness came upon her unbidden and unsought, often when she was walking or sitting in the open air and 'my heart dissolved in what I saw'. Ordinary things were transformed. Walking to the Sympsons after dinner, 'As I lay

down on the grass, I observed the glittering silver line on the
ridges of the Backs of the sheep . . . beautiful but with something of a strangeness, like animals of another kind—as if
belonging to a more splendid world'.[49] Sometimes the sense
was almost overwhelming; '. . . there was something in the air
that compelled me to serious thought. The hills were large,
closed in by the sky . . . the moon came out from behind a
mountain mass of black clouds—O the unutterable darkness of
the sky and the earth below the moon! and the glorious brightness of the moon itself!' It made her, she said in unaffected
modesty, 'more than half a poet' with 'many very exquisite
feelings'.[50] These were not, it is clear, worship of the natural
world itself, but the mysterious sense of a holy presence behind
what she saw. Later, at a time of deep emotion, returning
through Aysgarth after William's wedding, she found 'too
much water in the river for the beauty of the falls . . . a place
that did not in winter remind one of God. . . . There was something so wild and new in this feeling, knowing as we did in the
inner man that God alone had laid his hand upon it that I could
not help regretting the want . . .'.[51] The sense of God's handi-
work overwhelmed her one May evening of belated hail or
snow on the walk to fetch letters from Rydal. The woods seemed
half shrivelled with cold, and suddenly, 'O thought I! what
a beautiful thing God has made winter to be by stripping the
trees and letting us see their shapes and forms.'[52] At such sights,
belief came naturally as breath. In middle age, family griefs
and a changing social climate would lead her back to orthodox
worship, yet even in the Wordsworths' free-thinking years her
natural temperament seems religious.

Grasmere, with all its beauty, was isolated. They had not
seen a single new book since they came there and Dorothy confessed that 'William craves newspapers'. He was increasingly
troubled by the spread of manufacturing, which he felt must
weaken 'the bonds of domestic feeling among the poor'; here
lay one of the roots of his later conservatism. Friends and visitors
were a necessity. Both invited Charles Lamb in January 1801,
but he could not afford the journey, and added heretically,
'separate from the pleasure of your company, I don't much care
if I never see a mountain in my life'.[53] Sara Hutchinson came
in March and Mary promised to come in autumn. Meanwhile,

'do not fear dear Mary that you ever write too often'.[54] August 1801 brought Dorothy's childhood friends. Her schoolfellow Patty Ferguson, with Sam on a stay of some months from America, and another Halifax cousin, Mary Threlkeld, came on a visit which proved happy for all of them. They rowed on the lake and boiled a kettle on the Island for a picnic which Dorothy remembered whenever she saw a circle of smoky stones left by their fire. The memory acquired 'a tender melancholy' after the early death of 'that gentle creature Mary Threlkeld'.[55]

The friend of friends, Coleridge, was only a few miles away in his new home at Keswick. Yet though the Wordsworths stayed at Greta Hall, apparently four times in 1800 and 1801, Dorothy was distressed by the open unhappiness of his marriage. She attempted to be fair to Sara Coleridge, thinking her 'very much to be pitied, for when one party is ill-matched the other necessarily must be so too. She would have made a very good wife to many another man.'[56] Always willing to take on extra work, she had Hartley to stay for some weeks—'Dear little fellow, he is well and happy.' Coleridge walked over constantly, as in the Alfoxton days, but his visits were too often darkened by mysterious illness, with sufferings so intense that Dorothy trembled. In December 1800 he arrived soaked on the walk from Keswick and had to be nursed for two weeks. William believed his illness rheumatic fever or 'manifestly the gout',[57] since his joints ached and his fingers grew knotty. But he complained all through the winter and spring of 1800-1 of profuse sweats, 'bowels very bad', and 'dizzy head'. Dorothy wrote that he was 'dreadfully pale and weak . . . ill all over, back, and stomach and limbs'. She persuaded herself that a warm climate would cure him. 'We are sadly grieved for your poor eyes and the rest of your complaints,' she wrote to him in May 1801, 'but we sorrow not without hope.'[58] Mysteriously and slowly, hopes were to fail. She did not know that their brilliant, spellbinding friend was suffering the classic addict's symptoms of drug withdrawal. He struggled, bravely and with increasing self-disgust, but was forced back to reliance on opium. Dorothy grieved for his continual ailments; 'sad news—poor fellow!' she told Mary. 'I fear he has his own torments.'[59]

10. *'Either of joy or sorrow'*

This autumn 1801, Dorothy too began to show signs of recurrent illness. During the previous two years at Grasmere, her health, which had apparently survived the rigours of the German trip, was good. Long and strenuous walks in all weathers do not seem to have tired her or spoilt her extreme delight in everything she experienced. On 15 October 1801 however there was a change. On a walk up Loughrigg Fell, Dorothy held her head under a waterspout, and once home went to bed in the sitting-room, taking laudanum because she felt 'very sick and ill'. A month later melancholy letters from Coleridge kept her awake most of the night and she was not well enough to get up to breakfast. Next February she lay down 'very unwell' after reading letters which related the sufferings of Annette and little Caroline. A further letter from France left her sick with 'stupefaction and headache'. William had already noticed that, contrary to her former good health, she was 'subject to bilious sicknesses from time to time'.[1] On journeys, whatever the weather, she chose to ride in the open on top of the stage-coach, partly for cheapness' sake, but also apparently to avert travel sickness from the foul air inside. Her extreme activity, habitual overwork, vomiting, and other factors have recently suggested a diagnosis of migraine, and that she was 'a particular type of migrainous personality'.[2]

The attacks were accompanied by further troubles, which she described with Georgian forthrightness to Catherine Clarkson, her most intimate friend. 'I began, as usual with sickness, followed by a complaint in my Bowels with a violent looseness that lasted four days.'[3] Such attacks were severe enough to keep her in bed. Already thin, she lost weight, and William was 'rather alarmed' by the repeated shocks. The headaches improved with time, and there is no reference to them in letters after 1808; yet the painful spasms of the bowel continued, particularly in intense cold or at times of stress. After some years Dorothy accepted that 'I have fits of illness occasionally and probably shall have all my life'.[4] No attempt was made to diagnose these attacks until they became with old age a disabling illness. From early on though, Dorothy probably recognized the underlying causes better than anyone else. She became quite clear that 'any agitation of mind either of joy or sorrow will bring it on—if anything puts me past my sleep—for instance'. She defined these agitations even more precisely: 'any strong excitement cures my diseases for a time, but if I am well as surely brings them on'. It is noticeable that the year, autumn 1801 to autumn 1802, contained extreme joys and sorrows, often from events far beyond the control of anybody at Dove Cottage: even, indirectly from national events.

Private persons, of necessity, formed their plans and lived their lives within the framework of a nation at war. In summer 1801 gunboats and flat-bottomed barges were observed on the French coast, and on 21 July a secret circular to all District Commanders ordered 'the utmost vigilance to be observed throughout your district' against invasion of Napoleon's army. Yet already Nelson's guns at Copenhagen had broken the northern continental blockade, and Pitt, the great war leader, had resigned. On 1 October the preliminaries of peace with France were signed in London, and confirmed by Napoleon four days later. The formal Peace of Amiens was not to be concluded until 27 March 1802, but already mail coaches hung with laurels and banners reading 'Peace with France' sped from London to provincial towns, bringing public relief and joy. To Dove Cottage they brought a further, unwritten message; mail might once again come to England from France, from Annette.

This was the background against which Dorothy and William in November 1801 welcomed Mary Hutchinson to Grasmere for a seven weeks' stay. A shared life was nothing new to the three; they had lived contentedly together at Racedown in 1797 for seven months, at Sockburn in 1799 for another seven, and at Dove Cottage itself the previous spring for five weeks. Yet this winter visit was to make a new relationship. The two young women took up the homely routine of the household; baking bread and giblet pies, putting books in order, mending stockings, walking to the top of the hill on a sunshiny morning 'while the goose was roasting'.[5] Coleridge had departed to London for the winter on 10 November, and Dorothy anxious for him, 'eased my heart by weeping—nervous blubbering says William',[6] somewhat sternly. All the same, happiness grew in the cottage. William and Mary returned from a walk through a heavy storm 'chearful, blooming, and happy'.[7] Their love for each other could not be hidden, even from old Molly, who, Dorothy wrote, was 'very witty with Mary' about William's mature appearance. 'She says: "ye may say what ye will, but there's naething like a gay auld man for behaving weel to a young wife".'[8] It appears that during this November they reached a clear understanding about their future marriage. Dorothy walked with them, or sat with Mary on a beautiful frosty morning 'in our cloaks upon the bench in the orchard'. She visited neighbours with them, or shared evenings reading and writing letters by the fire. December brought heavy snow which buried grass and shrubs, leaving the ash trees, 'glittering spears with their upright stems'. William 'cleared a path to the necessary', and called Dorothy out to admire it, but before she got there a 'whole housetopful of snow' fell and blotted it out. At night the white fields shone in the moonlight.[9]

Grasmere, however, was none the less part of the outside world. On 21 December Mary walked through deep snow to the post office at Ambleside to fetch their letters. Among them was 'one from France'.[10] No plans for the future could be made without reference to Annette and to William's daughter Caroline, now nine years old. Dorothy had played a leading and personal part in the correspondence with Annette nine years earlier, before war had clamped down on communication. She had even tried to write 'all that is affectionate to A. and all that is fatherly to C.' for William during the war years. She was bound to be deeply involved now, with all the stress and strain on her nature and health.

For the moment she must see Mary on her way back to Durham. On 28 December 1801, William, Mary, and Dorothy, with 'some cold mutton in our pockets', set off on foot, the two women together and William part of the way alone, since he was 'busy in composition'. After a night at Keswick they pressed on through high wind and hailstorms towards the Clarksons' house, by an icy road. 'I was often obliged to crawl upon all fours and Mary fell many a time.'[11] They stayed three weeks with the hospitable Clarksons, during which Mary visited her relations in Penrith and then rejoined them. Mrs Clarkson, though unwell, entertained the company with tales of her two old aunts, 'Sister Barnard' and 'Sister Harmer', who lived together at Norwich. Sister Barnard kept two maids yet had 'a grand cleaning day' twice a week and whitewashed her kitchen herself. Dorothy wished she had set all the details of these stories down 'while they were fresh in my memory'.

On 23 January 1802, having parted from Mary, the Wordsworths set out to walk home, needing all William's skill to find the right track over the mountains through mists, with 'no footmark upon the snow either of man or beast'. It was dark when they reached Grasmere, thankful to change their wet clothes and sit by their own fireside.[12] On 26 January, they sat up late talking by the fire and 'William wrote to Annette'.[13] Two months of correspondence followed, which weighed on Dorothy's heart: on 16 February, after a letter from Annette, she 'went late to bed and slept badly'. A week later came letters from Annette and the small Caroline, when Dorothy 'lay down as soon as breakfast was over, very unwell'.[14] William answered them promptly, but on 22 March came another letter, perhaps with some account of the missing years—'poor Annette'. William, too, was sleepless, haunted by his work, anxiety, and as he later told Mary, by desire for her, 'those thoughts & wishes which used to keep sleep from me at Grasmere'.[15] Brother and sister on 22 March finally reached what seemed a joint decision. 'We resolved to see Annette and that Wm. should go to Mary', presumably to explain their proposed journey to France. Four days later, William wrote again to France, while Dorothy sat for two hours in the orchard. 'I was ill and in bad spirits.'[16] On 27 March 1802 the Peace of Amiens was signed and travel to France became a practical possibility.

On the morning of 28 March they made the first stage of William's journey to see Mary, arriving at Keswick 'wet to the skin'. They spent a week with Coleridge.[17] On 5 April the Wordsworths went on to Eusemere where Dorothy would stay with the Clarksons, while William went for a week's meeting with Mary at her younger brother George's farm, near Bishop Auckland in Durham. Dorothy walked with him six miles of the way upon his birthday, 7 April, and returned to wait for news. 'A windy morning—rough lake—sun shines—very cold —a windy night,' she wrote, turning to continuing realities.

On 12 April she collected a joint letter from William and Mary. A well-meaning neighbour walked with her, and questioned her 'like a catechizer. . . . Every question was like the snapping of a little thread about my heart—I was so full of thought of my half-read letter and other things. I was glad when he left me.' Alone, while the Clarksons were playing cards, she read of their agreement about Annette.[18] Next day, unexpectedly soon, William himself arrived. 'The surprise shot through me.' By 15 April they started on the two-days' walk home.

Their way lay beside Ullswater, rough in the storm. 'The wind seized our breath.' The deer in Gowbarrow deer park, after the long winter, were 'like skeletons', yet the birches were greening and there were a few primroses by the roadside. Then, in the woods near the water's edge, they saw the first daffodils, 'as we went along there were more and yet more; and at last, under the boughs of the trees, we saw that there was a long belt of them along the shore, about the breadth of a country turnpike road. I never saw daffodils so beautiful. They grew among the mossy stones about and about them; some rested their heads upon these stones as on a pillow for weariness; and the rest tossed and reeled and danced, and seemed as if they verily laughed with the wind, that blew upon them over the lake.'[19] Wind, rain, and the sound of waves like the sea could not blot out the daffodils. They slept the night at Patterdale, and walked home over Kirkstone next day in still, shining weather, arriving in 'half-moonlight, half-daylight' at their own garden.

The same evening, losing no time, Dorothy sat down at '¼ past 11 by the watch, but you know it is a little wrong headed —it is only ¼ past 10', to write to Mary. It was a letter glowing with generous affection. 'O Mary, my dear Sister!' she wrote for the first time, 'be quiet and happy . . . do not make loving us your business, but let your love of us make up the spirit of all the business you have.' She described the homeward journey. 'We sauntered and rested, loved all that we saw, each other and thee, our dear Mary.' The love found its resting-place in the cottage. 'Dear Mary, we are glad to be at home. No fireside is like this. Be chearful in the thought of coming to it. I long for a letter.' William added his own greeting at the end. 'Heaven bless you, dearest Mary.'[20] In the event, they did not marry for five and a half months. Problems still remained: poverty and dependence, the future of Annette and Caroline, all of which stirred Dorothy's deep personal feelings to the point of distress. Yet this letter, so transparently sincere, confirms that she loved Mary, both for William's sake and for her own. There was no suggestion that she should live apart from the pair; instead, a lifetime of loyalty through all trials united the two women. On the previous evening in the inn at Patterdale, she recorded, 'We had a glass of warm rum and water. We enjoyed ourselves, and wished for Mary.'[21]

Coleridge caused her more complex emotions. Returned from London, he had burst into the cottage on 19 March in a distraught state, and 'seemed half-stupefied', either from an encounter with his wife or from drug-taking. It is possible that Dorothy, who afterwards sat up late with William, now learned the full extent of his addiction. At all events, 'my spirits were agitated very much'. A month later, on 21 April, they were further distracted when Coleridge read her the long, despairing stanzas he had written to Sara Hutchinson; 'I was affected with them . . . in miserable spirits'.[22]

Yet, it seems, this time of stress on two fronts, the Mary-Annette and the Coleridge-Sara, proved a tremendous stimulus to perception and feeling. Though there are many references in Dorothy's journals not only to her illness but to William's (especially after frustrating efforts to write poems), the spring and early summer of 1802 found both Wordsworths at the height of their creative powers, William in verse, Dorothy in the prose of her journal entries. By 29 July, Coleridge calculated that William 'has written lately a number of Poems (32 in all) some of them of considerable Length'.[23] These, in fact, included major works such as the first four stanzas of the Immortality Ode, and the 'Leech-Gatherer' poem, *Resolution and Independence*. Very many of these poems were based either on experience shared with Dorothy, some years, or even several years, before (such as the glow-worm poem, a memory of Racedown),[24] or on incidents during the last few months. One, 'To A Butterfly', simply arises from their conversation. Though often quoted, it is worth repeating for the immediacy both of the poem and her description, and the fact that, as in the glow-worm poem, she is personally addressed by William's usual poetic name for her,

Emma, or in this instance, perhaps to fill up the line, Emmeline.*

> Stay near me—do not take thy flight!
> A little longer stay in sight!
> Much converse do I find in thee,
> Historian of my infancy!
> Float near me; do not yet depart!
> Dead times revive in thee:
> Thou bring'st, gay creature as thou art!
> A solemn image to my heart,
> My father's family!
>
> Oh! pleasant, pleasant were the days,
> The time, when in our childish plays,
> My sister Emmeline and I
> Together chased the butterfly!
> A very hunter did I rush
> Upon the prey: with leaps and springs
> I followed on from brake to bush;
> But she, God love her! feared to brush
> The dust from off its wings.

Dorothy's own words tell the genesis of the poem in simple, unforgettable detail:

While we were at Breakfast that is (for I had breakfasted) he, with his Basin of Broth before him untouched and a little plate of Bread and butter he wrote the Poem to a Butterfly! He ate not a morsel, nor put on his stockings, but sate with his shirt neck unbuttoned, and his waistcoat open while he did it. The thought first came upon him as we were talking about the pleasure we both always feel at the sight of a Butterfly. I told him that I used to chase them a little but that I was afraid of brushing the dust off their wings, and did not catch them . . .

The whole of the month of March is full of these domestic pictures. On the 23rd

The fire flutters and the watch ticks I hear nothing else save the Breathing of my Beloved and he now and then pushes his book forward and turns over a leaf.

A note on the 27th, announcing the Immortality Ode, is sometimes quoted as an instance of incongruity: it is rather an example of sheer practical naturalness.

---

* Perhaps counterparts of her two habitual family names, Dolly or Dorothy.

A divine morning. At Breakfast William wrote part of an ode. Mr. Olliff sent the dung and Wm went to work in the garden.

What Dorothy is above all describing is the atmosphere of a settled home, 'in our native mountains, there to live'.

Perhaps the key to these feelings lies in the lines William wrote after Dorothy's conversation about butterflies.

> A solemn image to my heart,
> My father's family!

Dorothy had now had a settled family home for over two years. Moreover, it was a home, though not in Cockermouth, in what might be regarded as Wordsworth country. Hence her remark, on 24 March, 'I made a vow that we would not leave this country for G[allow] Hill' in Yorkshire, even if Mary suggested it.[25] The Wordsworth family had all been deprived by the early deaths of their parents, but Dorothy far more than her brothers, because of her father's unaccountable failure to see her at all during the last half-dozen years of his life. Now she had re-created her 'father's family', in her father's part of the country, with William, older-looking than his age, as Molly the servant had observed, in some sense in a father's role. There can be no neat psychological formula for her state of mind, but the sense of security, even amidst external stress, can be felt everywhere in her journals at this time. Even depression over Coleridge, such as she felt at the time he recited the Sara Ode, did not last long, and could not compete with

Our own dear Grasmere first making a little round lake of nature's own with never a house never a green field but the copses and the bare hills enclosing it and the river flowing out of it. Above rose the Coniston Fells in their own shape and colour. Not Man's hills but all for themselves, the sky and the clouds and a few wild creatures.[26]

A hot beautiful day on 4 May contains almost all the chief elements in Dorothy's life at this time. First, she copied some stanzas of the Leech-Gatherer poem, which William had begun composing the night before, while he continued, in bed, to write several more stanzas. They took out a picnic lunch, and dinner, met Coleridge on the fells, and pressed on to a waterfall in 'a glorious wild solitude under that lofty purple crag. . . . A Bird at the top of the crags was flying round and round and looked in thinness and transparency, shape and motion like a moth.' All three had their picnic dinner by the river in the valley. 'Wm. and C. repeated and read verses. I drank a little Brandy and water and was in Heaven'[27] a remark which, strangely, is always quoted as an instance of naïve innocence. It is no more so than her 'glass of hot rum and water' less than three weeks earlier, and simply expresses another aspect of total enjoyment. They parted from Coleridge at Sara's Crag, where Dorothy, in another moment of spontaneous impulse, kissed all the carved initials.

Even such a perfect day would not have been typical if it had not contained some tax on Dorothy's ready sympathies. She met, walking, a deserted wife 'with 2 little girls in her arms the other about 4 years old walking by her side', half-starved and wearing ill-fitting slippers, cast off from some gentleman's child. 'Alas too young for such cares and such travels.' The mother 'was a Cockermouth woman 30 years of age—a child at Cockermouth when I was. I was moved and gave her a shilling —I believe 6d. more than I ought to have given.' Even North-Country thrift was not proof against this further reminder of her own lost childhood home and 'My father's family'.

William, in fact, was showing some of the less agreeable aspects of the conventional paterfamilias, especially on the subject of his own poetry. On 9 May, he had completed a draft of *Resolution and Independence*, and sent a copy to Mary and Sara Hutchinson, as virtual members of his new family. To his ill-concealed chagrin, both offered criticisms. On 14 June, he wrote them both letters,[28] which cannot hide his asperity, particularly on Sara's having remarked that she found some of the speeches of the Leech-Gatherer 'tedious'. Even Mary was sternly dealt with. While William was soothing his ruffled feelings by reading the poem to the visiting hair-cutter, below stairs, Dorothy herself felt obliged to add a special note to Sara in support of William—'ask yourself whether you have hit upon the real tendency and true moral . . . and when you feel any poem of his to be tedious, ask yourself in what spirit it was written'. The sequel is laughable enough. When the poem was printed five years later, all Sara's and Mary's criticisms had been followed, even down to individual words, and most of the 'tediousness' removed by abbreviation.[29] Dorothy's automatic rush to justify William was not only ridiculous in itself; it pointed the way to dangers which even Coleridge was quick to notice barely a year later. The 'perfect electrometer' he had praised in Dorothy's judgement and taste was now switched off as far as William was concerned. Fortunately, Dorothy's relations with her 'Beloved Friends' Mary and Sara were usually much happier than this, as affectionate letters show.

The months before the journey to see Annette, and William's marriage, were a time of practical decisions. Coleridge hoped that the new household would live or lodge with him at Greta Hall, but Dorothy knew this would be unwelcome to his wife and refused firmly: 'I said I could not see any good whatever to arise from this, and as I was so fully determined he pressed nothing upon me.'[30] Instead, she worked to make the cottage and garden ready for Mary, reporting that Molly 'looked clean and handsome and the house is a perfect model of neatness'. The previous October, Mary's brother Tom had given her—rare extravagance—'2 shrubs from Mr. Curwen's nursery';[31] in January she pruned, and on 2 February 1802 Charles Lloyd brought her flower seeds. In April they dug and planted. The first day of May was 'A heavenly morning. As soon as breakfast was over went into the garden and sowed the scarlet beans about the house. . . . I sowed the flowers and William helped me.' That week they planted 'our Bower' for sitting outdoors and on 8 May William added a step to the steep orchard stairway. She planted wild daisies, foxgloves, and lilies 'about the orchard' and flowering turf 'about the well'. By 28 May 'the scarlet beans are up in crowds'; they hoed and worried about rain. By the last week of June 'John's rose tree is very beautiful blended with the honeysuckle'. On 6 July, although snow had fallen on the mountain tops, 'the well is beautiful. The orchard full of foxgloves, the honeysuckle beautiful, plenty of roses but they are battered.' Meanwhile she had found time to put the finishing touches to the house, helped by the energetic clergy-daughter Margaret Sympson, in a traditional country spring-cleaning, which included painting. On 25 June 1802 'Miss Sympson came to colour the rooms. I began with white-washing the ceiling.' House and garden, by her devoted labour, stood ready to welcome their new mistress.

Less agreeable were the necessary discussions about her own

means of support. She assured Richard that Mary was trustworthy: 'I have known her long and I know her thoroughly',
she wrote on 10 June. 'She has been a dear friend of mine, is
deeply attached to William, and is disposed to feel kindly
to all his family. . . . I shall continue to live with my Brother
William.' She dismissed any thought of her own marriage
as 'absurd at my age (30 years)'. In answer to Richard's
question about a settlement she said, 'Sixty pounds a year is
the sum which would entirely gratify all my desires.' It was
also, incidentally, the highest income free of the Income Tax
imposed by Pitt in 1799 to pay for the French Wars. Of this
Dorothy already received twenty pounds a year 'absent money'
from John's East India Company salary, and twenty pounds
a year from Christopher out of his stipend as a Fellow of Trinity,
Cambridge. William could not contribute, 'he having nothing
to spare nor being likely to have. . . . I am obliged (I need not
say how much he regrets this necessity) to set him aside.'
Characteristically, Dorothy took on herself the task of breaking
this unwelcome news, in order to spare him. She did not like to
beg from Richard, 'I should be very loth to be oppressive to you
or any of my Brothers, or to draw on you for more than you
could spare without straitening yourselves', but relied 'as I do
and have ever had reason to rely' upon their affection.[32] The
letter spells out what William called her 'utter destitution'
through the Lowther lawsuit, and her dependence upon charity.
Perhaps writing it, together with hearing again from Annette,
left her on 14 June 'very unwell—went to bed . . . was sick' with
'stupefaction and headache'.[33]

Any agitation, for good or ill, had its instant effect on Dorothy's health. Four days later, she again 'had a woful
headache, and was ill in the stomach'. This time, the cause was
a joyful one. Their old legal enemy, Lord Lowther, had died on
24 May, and on 18 June they received the news that his heir
would pay all just debts. Optimistic as ever, after her first sickness from shock, Dorothy with William soon 'talked sweetly
together about the disposal of our riches',[34] though, in fact,
legal business delayed any payment of the Lowther debt until
the following year. It was not the Lowther money which made
William's marriage possible, but Dorothy's earlier decision to
boai d as with 'an indifferent person', paying her share of his

household expenses.[35] The Ferguson cousins, always concerned for her, expected 'something Handsome' to 'make Dorothy quite comfortable' from the Lowther estate, yet her everyday life changed little. At thirty her character was long set in habits of hard work, frugality, and unquestioning self-sacrifice.

In the first week of July 1802, she prepared for the journey with William to France. There was a final exchange of letters with Annette, one shown or sent to Mary. Dorothy ironed the linen, did the packing, wrote to John, said her silent farewell to home. 'Glow worms out, but not so numerous as last night. O, beautiful place! Dear Mary, William. The horse is come—Friday morning—so I must give over.'[36] They spent two nights at Keswick with the Coleridges, two at Eusemere with the Clarksons, and took the coach from Emont Bridge into Yorkshire for a final visit to Mary and the Hutchinsons, now farming at Gallow Hill, near Malton. On the sixty-mile coach drive a storm came on, but they buttoned themselves together into the guard's cloak and braved it; 'I never rode more snugly'. The last stage, over the Hambleton Hills, they walked, Dorothy thirsty and footsore, but rewarded by looking down on the ruins of Rievaulx Abbey 'among a brotherhood of valleys'.[37] From 16 to 26 July they stayed with the Hutchinsons, exploring the Wolds in bad weather, with gleams of sunshine. Sara was with her sister, which was fortunate, for even unworldly Dorothy Wordsworth knew that her Grasmere clothes would hardly do for France, the centre of European fashion. Sara had views on clothing; she had ordered for herself from London 'a Chip Hat or bonnet of the very *newest* fashion . . . pea-green else lilac'.[38] Moreover, unlike tall Mary, she was as small as Dorothy herself. She lent one or more 'white gowns' and Dorothy felt she could 'trust to Sara's wardrobe'.[39]

By 29 July Dorothy and William were in London, and two days later at half-past five or six in the morning they mounted the Dover coach at Charing Cross for a journey which between them they immortalized. 'It was a beautiful morning,' wrote Dorothy. 'The city, St. Paul's, with the river and a multitude of little boats, made a most beautiful sight as we crossed Westminster Bridge. The houses were not overhung by their cloud of smoke, and they were spread out endlessly, yet the sun shone so brightly with such a fierce light, that there was even some-

thing like the purity of one of nature's own grand spectacles'[40]
She was going to see the unknown Annette, whom nine years
earlier she had welcomed as a sister, and to whom she had written so faithfully; more moving still, she was going to see William's child. For both Wordsworths the sleeping city cradled
deep human feelings. The famous sonnet, which William probably began now, and perhaps finished on his return to England
—'on the roof of a coach' he afterwards said—has an agitation
of feeling and an intensity which speaks of deep inward associations with the occasion, a crucial moment in the isolated hearts
of himself and Dorothy.

> Dear God! The very houses seem asleep;
> And all that mighty heart is lying still!

On 1 August 1802 they were still on board in Calais harbour
at half-past seven in the morning, but by about half-past eight
they had 'found out Annette and C. chez Madame Avril dans
la Rue de la Tete d'or.'[41] They were to spend the whole month
together, a stay some writers have found surprising, but it was
a time of decision in all their lives, too serious to be hurried.
Physical change alone demanded acceptance. Was Annette's
once-ardent lover this serious, rather shabby figure? Was this
thin, work-worn woman his enthusiastic young sister? Annette
herself was now thirty-six, for years a fighter heart and soul in
the Catholic and Royalist resistance party known as the
Chouans. She had taken the name Madame or Veuve Williams
and brought up Caroline by herself, the troubles of a single
mother swallowed up in the conflicts of France. With her two
sisters she had hidden in the family house at Blois an unknown
number of priests and royalist refugees of all classes, one of
whom later declared that 'she had saved his life by risking her
own'. She attended secret Catholic services and is said to have
arranged a prison rescue by rope-ladder. She appeared in police
records as 'Widow Williams at Blois; gives shelter to the
Chouans' and had lived in daily danger of arrest.[42] Perhaps the
strongest emotion she still shared with William was hatred of
Bonaparte.

Dorothy was deeply moved by this history of suffering and
danger. Whatever her own suspicions of France and the French,
and later, following William, of Roman Catholics generally,

she did not lose her affection and sympathy for 'poor, dear Annette', or her concern for Caroline. She loved most children, William's children above all. Now they walked almost every evening upon Calais Sands, 'delightful walks after the heat of the day was passed away'; beside them the 'dear child', a lively, inescapably French little girl, 'untouched by solemn thought', who yet grew to resemble William strikingly[43] and was also inescapably her niece. It was not Dorothy Wordsworth's way to write directly of her inmost feelings, which overflowed in an evocation of the summer nights. 'The reflections in the water were more beautiful than the sky itself, purple waves brighter than precious stones, for ever melting away upon the sands.' On hot, still, cloudy nights she saw 'the little boats row out of the harbour with wings of fire, and the sail boats with the fiery track which they cut as they went along'. Even here she saw with the child's eyes the 'sparkles, balls, shootings and streams of glow-worm light. Caroline was delighted.'[44]

A month of walks and talks confirmed the gulf ten years had opened between Caroline's parents, which both wisely recognized. Friendship and respect still remained between them, with a shared concern for their daughter's future; they discussed a dowry, but they met only twice more in their lives. Dorothy continued through the hazards of renewed war to write steadily to Annette, foreign, her country's enemy, yet admitted to the sacred circle of her family love. On 29 August the Wordsworths left Calais, and reached Dover harbour after twenty-five hours. 'I was sick all the way.' The last day of August was very hot. They both bathed, then 'Sate upon the Dover Cliffs, and looked upon France with many a melancholy and tender thought'.[45] To find after so long, and lose so quickly, a child of their own flesh and blood, had touched the depths of Dorothy Wordsworth's nature.

London brought immediate distractions, the stoical eighteenth-century remedy in which she had been reared. They arrived at six in the morning on 1 September and stayed for three weeks, lodging in the Inner Temple, probably in Basil Montagu's chambers, unused during the law vacation. Nearby, in Mitre Court Buildings, lived Charles and Mary Lamb; they invited the Wordsworths to dine on 7 September and Lamb, a Londoner heart and soul, showed them round 'Bartlemy Fair' as a treat.[46] Dorothy, who always got a headache from noise and crowds, found more happiness in a growing admiration for kind, generous, though afflicted, Mary Lamb. Instinctively she understood the tragedy of the devoted brother and sister, and the strength of the tie which bound them. Lamb had been caustic about Wordsworth's response to the mildest criticism, 'a long letter of four sweating pages from my Reluctant Letter Writer'.[47] Yet the sympathy of the two women led their families into deep and lasting friendship, proved when the Wordsworths, in their turn, suffered a family tragedy.

London also brought a rare concentration of family meetings, always occasions of deep emotion to Dorothy Wordsworth. Richard and William held business discussions, recently a source of acrimony between them.[48] Christopher, visiting from Cambridge, arrived at Richard's chambers in Staple Inn to meet them with good news. John, aboard the *Earl of Abergavenny*, had safely returned from his China voyage, but was detained off the Kent coast by neap tides. The other three went down to Windsor, to visit Canon and Mrs Cookson, the first meeting since Dorothy's contrived departure from Forncett eight years earlier. All was generously forgiven; she was touched to find her aunt little changed, uncle 'scarcely older', and the babies 'fine-tempered children', who welcomed her affectionately. The best was to come. Returning to London, by open coach, she entered the Temple Courtyard and saw John, walking back and forth with Richard 'by the light of the moon and the lamps'. She hurried him in and lit the candles to make sure he was truly safe and well. She and William ought to leave for Yorkshire, but, as she wrote to Mary, 'we cannot find it in our hearts to leave him till we have been a few days together'.[49] The Wordsworth men, so unexpectedly gathered together, were varied: cautious lawyer, correct young clergyman, silent and shy sea-captain, with one uncompromising poet suspended between his past and future; yet to Dorothy they were her brothers, beloved above all common humanity. 'It was', she wrote, 'a delightful thing thus to see all our Brothers.' The mounting joys and sorrows of the past weeks overcame her. She had caught 'violent cold' in the coach from Windsor and become 'exceedingly unwell', with an attack of her familiar symptoms this time more serious than ever before. She fell ill on her return from the

Cooksons about 14 September, and was too weak to set out on
the coach journey into Yorkshire until 22 September. They
arrived, after travelling day and night, at Gallow Hill two days
later, and Mary, 'in rich and flourishing mien' as her bridegroom remembered, came down the lane to meet them. She
was shocked by Dorothy's thin and exhausted looks. The garden
was 'gay with asters and sweetpeas', farmer Tom stood on his
cart forking corn, and the three Hutchinson sisters took such
good care of the invalid that she hoped to grow strong. Yet, in
spite of herself, she remained 'ill on Saturday and Sunday and
continued to be during most of the time of our stay'.

She welcomed, yet dreaded, her brother's wedding, as she
explained in a letter on 29 September to Jane Marshall, which
bears the stamp of simple, habitual truth. 'I have long loved
Mary Hutchinson as a Sister, and she is equally attached to
me'; fifty-three years of shared life would bear out the truth of
this. 'This being so,' Dorothy continued, 'you will guess that I
look forward with perfect happiness to this Connection between
us, but, happy as I am, I half dread that concentration of all
tender feelings, past, present and future which will come upon
me on the wedding morning.' An experience in her life was
ending and she felt too frail for strong emotion; she wished only
'the wedding was over' and she could be at 'our own dear
Grasmere with my most beloved Brother and his Wife'.[50] Five
days later she was not calm or well enough to leave the house
for the church at Brompton-on-Swale. Of all the large Wordsworth and Hutchinson families, only Tom and John as legal
witnesses, and Joanna as bridesmaid, were present at eight
o'clock on Monday morning, 4 October 1802 for the marriage
of William and Mary Wordsworth. Dorothy watched from a
window and saw them go down the avenue towards the church.
Her journal records:

I gave him the wedding ring—with how deep a blessing! I took
it from my forefinger where I had worn it the whole of the night
before—he slipped it again onto my finger and blessed me
fervently.[51] When they were absent my dear little Sara prepared the
breakfast. I kept myself as quiet as I could, but when I saw the two
men running up the walk, coming to tell us it was over, I could stand
it no longer and threw myself on the bed, where I lay in stillness,
neither hearing nor seeing anything till Sara came upstairs to me and
said 'They are coming'. This forced me from the bed where I lay and

I moved I know not how straight forward, faster than my strength could carry me till I met my beloved William and fell upon his bosom. He and John Hutchinson led me to the house and there I stayed to welcome my dear Mary.[52]

No other words could match the transparent truthfulness of this account. Mary, accepting Dorothy's feelings with the rare generosity which marked her whole life, walked home, not with her own bridegroom, but beside her bach lor brother Tom with whom she had lived for years.

The wedding breakfast at the farmhouse was a simple family affair, no speeches, no ceremonies, no guests, and no wedding presents, except for a new gown for Mary from John Words-worth, the shy bachelor sailor who had always been devoted, perhaps even a little in love with her. Dorothy herself received a handsome present towards the joint household, 'half a dozen silver table spoons' from her cousin Sam Ferguson who was flourishing as a merchant in Philadelphia. He had heard William was to marry 'Miss Hutchinson, Dorothy's intimate friend,'[53] and took it for granted that they would all live together, since such family arrangements were common. On her side, Dorothy accepted the whole clan of Hutchinsons and their Monkhouse cousins into the Cookson and Wordsworth circle, a plethora of Thomases, Henrys, Georges, Johns, Williams, with three Dorothys, five Saras, and eight Marys. Her letters chronicle the affairs of some forty members of this tribe, happy and affec-tionate together, but in the end restrictive in their narrow family interests.[54]

As the breakfast drew to an end, 'Poor Mary was much agitated,' wrote Dorothy compassionately, at parting from brothers, sisters, and home. The Hutchinsons, like the Words-worths, had been orphan children; in both families brothers and sisters were all in all to each other. As soon as breakfast was over, they set off in rain, which yielded to 'sunshine and showers, pleasant talk, love and chearfulness.'[55] The only am-biguous comment on the new household came from Coleridge, away in Cumberland. He noted that he dreamt on 3 October of William, Dorothy, and Mary; in his dream 'Dorothy was altered in every feature—fat, thick-limbed & rather red-haired —in short no resemblance to her at all, and I said "if I did not *know* you to be Dorothy, I never should *suppose* it".'[56]

11. 'My tears will flow'

The three Wordsworths made their way home by coach through Wensleydale, where the river glittered in the sunshine, and Dorothy's heart 'melted away with dear recollections' of her walk with William over the same road to Grasmere three years earlier. Reaching the cottage after dark on 6 October 1802, they all went into the garden with candles, impatient to inspect the plants. Next morning the two women began the essential work of unpacking and bread-making with, later in the week, special baking of cakes. On 17 October they had thirteen of their cottage neighbours to tea, a homely gathering, where William appeared at the last moment, to celebrate Mary's official arrival as mistress of the household. In an interval between tasks, the young women walked on the hill looking towards Rydal; it was a special occasion, 'the first walk', Dorothy wrote, 'that I had taken with my Sister'.[1]

The sisterhood was an equal partnership. Both felt themselves essential to William, to his work, and later, overwhelmingly, to his children. A perceptive later visitor noted how Dorothy brought to mundane tasks the fire and ardour of an impassioned temperament, and Mary 'a sweetness almost unexampled of temper'. Dorothy and William might sometimes be heard to disagree, but Dorothy and Mary never: 'Impossible for anybody to have quarrelled with *her*'.[2] Mary had seen nothing of the world outside the small town of Penrith and her brothers' farmhouses. She made no claims to learning, accomplishments, or beauty, with what she later called her 'squinty eye',[3] and her manner of unaffected country simplicity. Her character proved strong, generous, stable under any stress, and her judgement excellent; to the passionate joys and sorrows of the Wordsworth temperament, she provided a calm counterpoise.

Mary and Dorothy made one complaint; each found the other too self-sacrificing. Dorothy felt 'exceedingly uneasy and anxious' that Mary worked so hard when suckling her children; 'if we were to suffer her to go on after her own inclinations she would in a very few years be worn out'.[4] Mary regretted that William's sister should be 'such a Slave. . . . I do not know', she wrote, 'that it is good for the heart at all times to be strict to the dictates of duty.'[5] Apart from this, their writings record decades—on Mary's side over fifty years—of unbroken love and loyalty. Relationships só deep do not grow of their own accord; they demand self-discipline on both sides. As in any family, there were danger points. William's letters to Mary make this plain. He felt 'the blessed bond that binds husband & wife so much closer than the bond of Brotherhood—however dear'. The difference lay in his intense physical passion; when he was separated from his wife, a 'fever of thought and longing and affection and desire is strengthening in me. . . . I tremble with sensations that almost overpower me.'[6] From this, the heart of their marriage, Dorothy was necessarily excluded. There is evidence in the same correspondence that she felt this painfully, and William knew it. When, by 'a most unlucky oversight', he wrote some loving passages to Mary on part of a letter which Dorothy might see, he could not send it without blotting out what he was afraid Dorothy might have thought were 'obnoxious expressions'.[7]

If the overtones of what William and Mary wrote to one another might be 'obnoxious' to Dorothy, one must wonder how she faced sexual relations between the newly married pair in such a small house. William, it is true, had already moved into the downstairs bedroom, and slept with Mary there, while Dorothy had her room upstairs: but, as she herself had written, 'from its smallness and the manner in which it is built noises pass from one part of the house to the other'.[8] She is impenetrably silent, only remarking on her birthday, Christmas Day, 'William and Mary are very well. I am quite recovered from my late illness.' Since the second week in November, she had had a companion, and presumably bedfellow, in Mary's sister Sara, who had visited her aunt Elizabeth Hutchinson at Penrith, on which occasion Coleridge had snatched the chance to spend a day there with her. Sara, racy, gossipy, and matter-of-fact, was a cheerful companion for Dorothy at this time; she stayed two months in the cottage, and 'we have had scarcely any company . . . since we came home, a mighty comfort!'[9] Coleridge had reappeared from a West-Country tour on the morning of Christmas Eve, inevitably attracted by Sara, but William and Dorothy greeted him with the news that his wife had borne him 'a fine Girl' the day before, and he left almost at once for Keswick, as a matter of duty. Coleridge's passion for Sara Hutchinson and growing estrangement from his wife were an unvarying theme all through the coming year. So was his self-confessed failure to work. 'My very heart dies!—this year has been one painful dream / I have done nothing!'

All through 1803, Coleridge and her brother John were Dorothy's twin poles of anxiety. Admittedly the latter, after his first voyage in charge of the *Abergavenny*, seemed at a peak of success. His financial investment in that ship's China voyage, as allowed by the Company, brought him more than he had anticipated when Dorothy last saw him in London, just before William's wedding.[10] He was now preparing for another profitable Far East voyage, on which Coleridge for a time applied to accompany him, hopefully deluding himself that all he needed was warm climate. 'I think it would be better if he would go to Italy,' wrote John tactfully to Dorothy.[11] In addition, there were prospects of danger as well as profit by sea, since the short-lived Peace of Amiens collapsed in May, and merchant vessels might suffer French attack, on the renewal of war.

The *Abergavenny* sailed on 6 May, without Coleridge, as might have been expected, but with an investment of some of Dorothy's share of the Lowther money, in spite of her own misgivings about 'the risque of losing this money by which means I might forfeit my independence'.[12] She regretted it was 'a very painful thing' to oppose Richard and John—who had apparently invested her money without first consulting her—but insisted that, in the unlikely event of loss on the voyage, her brother Richard should see she was reimbursed adequately, and, in any case, would pay interest on her capital while it was carried on this overseas venture.[13]

Anxieties for Coleridge and for John were soon swamped, however, by what became for her a ruling obsession, the first child born to William and Mary. Mary had conceived at once, and by the second half of January 1803 was nearly four months pregnant. Sara had gone home to Yorkshire on 7 January, and Dorothy, single-handed, took the heavy home burdens off the expectant mother. It is often asked why Dorothy ceased to keep her Grasmere journal after 16 January, in spite of determining to start a new book and 'write regularly and, if I can, legibly'.[14] Some unlikely reasons have been put forward; the most likely is that now, in spite of her resolve, she simply did not have the time. To support 'my Sister' in the second half of her pregnancy was a prospect she embraced with the whole of her loving feelings.

Nor did those feelings in any way diminish when, on 18 June, John Wordsworth, named after his uncle, was born, 'a stout healthy Boy'. Dorothy's expressions of joy became, and remained, ecstatic. 'Oh my dear Friend,' she wrote to Mrs Clarkson, 'how happy we are in this blessed Infant!' The nurse was sent away only three days after the birth—'Mother and Child have gone on so nicely. I have been their sole attendant.' The praises of Johnny redouble after his christening on 17 July, with herself as godmother, Coleridge in person and Richard (by proxy) as godfathers. In a cradle improvised from a meat basket, he went everywhere with his adoring family, 'a noble looking Child', and though after a few months Dorothy had to admit he was 'not a great *Beauty*', yet she reiterated, he was 'a noble looking creature'. It was clear there could be no writing of journals while she was carrying Johnny around in her arms. Even writing letters was difficult. 'The darling is sitting upon my knee—how I wish you could see him with his serious face tracing my pen, and how his bonny eyes are lighted up by a smile, and now the little darling sends out his voice into every corner of the house.'[15] Johnny's 'angry squawls' when teething and 'violent passion' when being dressed were meekly accepted by his loving aunt; 'what peace and pleasure, wakefulness and hope there is in attending upon a healthy infant . . . one's thoughts are never tired when so employed'.[16]

She was filling the traditional role of maiden aunt, honoured and happy in so many families. Mary Lamb, a lonely spinster, wrote wistfully, 'Are you not now the happiest family in the world? . . . When, and where, shall I ever see you again? Not I fear for a very long time, you are too happy ever to wish to come to London.'[17] Cassandra and Jane Austen found endless pleasure in their brothers' numerous children: 'dearest Fanny', her 'agreable idle brothers', 'little Harriet', and others. Yet Jane Austen's cool eye noticed 'that puss Cassy' who '*ought* to be a very nice child . . . if they will only exert themselves a little', while later a niece too often pregnant was dismissed as 'Poor Animal, she will be worn out'. Dorothy Wordsworth was incapable of such detachment. Her adoration of Johnny poured out in language more overwrought than her schoolgirl passion for Jane, or even her devotion to 'my Beloved' William. The occasion was so simple and natural, yet there is something almost disturbing in the intensity of emotion it aroused.

De Quincey, who first met Dorothy when she was still obsessed by Johnny, by then aged four, later noted the harm that could be done to a doting aunt. In a passage clearly referring to Dorothy and her brother's oldest child, he afterwards wrote:

Nephews and nieces, whilst young and innocent, are as good almost as sons and daughters to a fervid and loving heart that has carried them in her arms from the hour they were born. But, after a nephew has grown into a huge bulk of a man, six foot high, and as stout as a bullock . . . there is nothing in such a subject to rouse the flagging pulses of the heart, and to sustain a fervid spirit. . . .[18]

This is almost exactly what happened to Dorothy and Johnny. Moreover, the 'noble forehead', so often remarked by Dorothy, proved far from a sign of intellect. Yet all this was hidden in the future. He was still 'the dear Babe' and 'as fine a boy as ever was beheld'. She was convinced that the first thing a new visitor to Grasmere would want to see was Johnny,[19] 'for I am sure the very sight of him would give you as great a delight as you have ever felt from any thing not known and loved before'.

It seems all the stranger, then, that exactly two months after Johnny's birth, she allowed herself to desert him and Grasmere for as long as six weeks. She did not depart without a pang. 'I do not love to think of leaving home, and parting with the dear Babe who will be no more the same Babe when we return.'[20] Conflicting emotions, though Coleridge thought it the 'hurry & bustle of packing', caused 'one of her bilious Attacks' as they were setting off, and delayed them for three days. It is not clear why on 15 August she set off for a tour of Scotland with William and Coleridge, nor, indeed, what was the reason for the whole expedition. Only one of them, William, had been to Scotland before, briefly, for Montagu's second marriage. It seems that the motive may have been a search for health, especially William's. 'William's health was very much amended by our Tour,' Dorothy reported, as if that were the main reason.[21] William himself said it had relieved 'a set of painful and uneasy sensations which [I had] more or less at all times about my chest',[22] and he thought Dorothy too was improved in health, though she already had put down an improvement to taking wine regularly, in spite of the expense. All the same, the choice of such a venture, comparatively late in the year, remains something of a mystery, particularly for Coleridge, who was already seeking a warm climate, which he notably failed to find north of the Border at this autumnal season.

It may be that 'the Concern' of William, Dorothy, Coleridge had hopes of renewing the enchanted atmosphere of Alfoxton days in a Northern setting. If so, they were disappointed. Just as the 'German tour', which followed Alfoxton, broke up after only a fortnight, so did the Scottish tour after the same short time, with Coleridge making his way alone, while Dorothy and William pushed on together for another month. The ostensible cause was the rigour of travel with an Irish jaunting-car and a nervous horse which William was singularly ill-adapted to manage. As Coleridge wrote, 'a Horse and Jaunting Car is *an anxiety*—and almost wish we had adopted our first thought and *walked*', with an ordinary pony and side-saddle for Dorothy. In fact, he proved his own theory, after parting from the two others at Loch Lomond, by achieving the huge loop of a Highland journey of 263 miles on foot in eight days. William and Dorothy went on west and north for another four weeks, deep into the Highlands, to Inverary, the Vale of Glencoe, Killiecrankie, and back via the Trossachs, where 'a laugh was on every face', wrote Dorothy 'when William said we were come to see the Trossachs'. They finished by Edinburgh, meeting Walter Scott, and home through the vales of Tweed, Teviot, and Esk, arriving back at Grasmere on 25 September. 'We had a joyful meeting' there with Mary, her sister Joanna, who had kept her company, and, of course, Johnny.

It had been an exacting journey. Dorothy, who had not been in good health at the start, was too exhausted each evening to do more than ensure they had adequate beds for the night and some food, a task which usually fell to her. She somehow persuaded reluctant or obstructive hostesses to 'give fire' at the cold inns where they mostly stopped, though she could do little about the inedible food, still less the generally prevailing dirt, 'quite Hottentotish'. Even with the dubious jaunting-car, she was continually soaked, sometimes more than once a day, and was often unable to borrow a temporary change of clothing. Yet an almost manic energy carried her through the six weeks, in a country which was too much for the athletic John Keats, fifteen years later, when roads and inns had improved. She was hardly helped by the accident-prone William, who was not only at odds with the horse, but who, on one occasion, dropped the whole of a day's picnic-food into the waters of a loch[23] She herself confessed to Mrs Clarkson that 'I was not as well as I could have wished at my return, but', she added, 'I now am perfectly well though I had a bad attack on Friday of my old complaint'.[24] It seems almost a miracle she felt no worse. Characteristically she at once began worrying not about her own health, but that of Joanna Hutchinson, who had found the strain of helping Mary at Dove Cottage too much, and had 'many symptoms of nervous diseases since she came to Grasmere', culminating in 'a hysteric and fainting fit'. Dorothy kept Joanna at the cottage for another week, and took charge of nursing her.[25]

Neither this nor innumerable household and family tasks, however, prevented her from doing what she had been too weary to do in Scotland, writing an account of the whole journey, a private recital for favoured friends such as the Clarksons. This is the more amazing since she had taken no notes of any sort. 'I am writing', she announced, 'not a journal, *for we took no notes*, but *recollections* of our tour in the form of a journal.'[26] She started this almost as soon as she reached home, and worked steadily until 20 December, when she was interrupted by the arrival, for a month's stay, of Coleridge, ill, exacting, and about to seek his desired warm climate abroad. She resumed work on 2 February 1804, but was checked again by another task for Coleridge, making a copy of all William's poems to take on his travels. Thrown out, she made fruitless efforts to continue, but, urged on by William, only went on again in April 1805, and finished on the last day of May in that year.[27]

This was, apart from all else, an astonishing feat of memory. It leads one to believe that she shared the extraordinary power of recall by which William could summon up his past life and remember, verbatim, hundreds of lines of recently composed poetry. Only by a powerful gift for re-creating places, incidents, people, and words can she have constructed this account, so like an on-the-spot journal that many have refused to believe it was not actually written during the tour itself.

For instance, while Coleridge merely reported that on Saturday, 27 August, near Loch Achray, Dorothy had a bed in 'the Hovel', a building with 'smoke-varnished Rafters',[28] her remembered account has the magic of earlier journals.[29]

I went to bed some time before the family. The door was shut between us, and they had a bright fire, which I could not see; but the light it sent up among the varnished rafters and beams, which crossed each other in almost as intricate and fantastic a manner as I have seen the under-boughs of a large beech-tree withered by the depth of the shade above, produced the most beautiful effect that can be conceived. It was like what I should suppose an underground cave or temple to be, with a dripping or moist roof, and the moonlight entering in upon it by some means or other, and yet the colours were more iike the colours of melted gems. I lay looking up till the light of the fire faded away, and the man and his wife and child had crept into their bed at the other end of the room. I did not sleep much, but passed a comfortable night for my bed, though hard, was warm and clean: the unusualness of my situation prevented me from sleeping. . . . I was less occupied by remembrance of the Trossachs, beautiful as they were, than the vision of the Highland hut, which I could not get out of my head.

The *Recollections* were, of course, written for an audience of close friends and relatives, for them both to read and to copy. Mrs Cookson made one such copy, and another was made by Sara Hutchinson, for Coleridge to read when he returned from abroad. Dorothy herself had no thought of publication. However, in 1820, the poet Samuel Rogers, who had read one of the manuscript versions, urged her to publish. Rogers had met the Wordsworths themselves early in the tour, when Coleridge was still with them, and when, he commented,

Wordsworth and Coleridge were entirely occupied in talking about poetry: and the whole care of looking out for cottages where they might get refreshment and pass the night, as well as seeing their poor horse fed and littered, devolved upon Miss Wordsworth. She was a most delightful person,—so full of talent, so simple-minded, and so modest![30]

Rogers's scheme for publication, in fact, fell through, and the early manuscript *Recollections* were not published until nearly twenty years after Dorothy's death. This was perhaps lucky, since the fair copy she rewrote for the press in the 1820s was very much revised in the interests of supposed respectability. All references to William's general clumsiness, for instance, were carefully removed for this intended publication.

This revision gives point to the feeling that the *Recollections*, though highly praised by lovers of her work—'undoubtedly, her masterpiece',[31] as one of them considered—do not have quite the magic of the more fragmentary Grasmere Journals, written generally on the instant, or, if delayed, only a few days after. The Scottish *Recollections*, fine as they are in description and atmosphere, do not have the limpid freshness of the *Journals*. They seem, in the slightest degree, over-considered. They do not quite breathe the same air of immediate, spontaneous joy and wonder. For one thing, it is noticeable that Dorothy frequently does not record her impressions of new scenery or incident, mountain, lake, glen, or settlement, for their own sake, catching their own unique quality. She has a minor but persistent habit of comparing these fresh sights with others she has known in different parts of England. They are not enjoyed for themselves, but because they resemble Skiddaw, Derwent Water, Ullswater, Grisedale, Grasmere, even places she had known in Dorset and Somerset: one Scottish mountain even reminds her of Glastonbury Tor, which she had mistakenly thought she saw from Alfoxton.[32] Reaching home when the tour ended, she was continually asked, 'Do you like the Scotch or the English Lakes better?' She replied that though there was no 'comparison to be made where everything is so different . . . there is nothing so *beautiful* in Scotland as parts of this country'.[33] Likewise in Germany, she had been limited by her overpowering attachment to home scenes and associations.

The second half of 1803 brought a double shock, which even the bliss of Johnny's company could not muffle. Catherine Clarkson, Dorothy's closest friend in the Lake Country, had a serious illness. Taking her little boy she left Ullswater for her native East Anglia, and visited Clifton to consult the fashionable Dr Beddoes. Dorothy asked Catherine to consult Dr Beddoes on her own behalf, since the 'complaint in my Bowels' had recurred for two years, and he obligingly prescribed without even seeing his new patient. Not surprisingly, his remedies proved ineffective. Meanwhile Catherine finally settled with

her family in Bury St. Edmunds. Both the Clarksons had their
roots in East Anglia and wanted to live, not shut in by mountains but under limitless skies, among their own people. Dorothy
Wordsworth could never wholly bring herself to accept this.
For her, Eusemere was their home, and Catherine must wish to
return 'every summer'. In May 1804 she walked over to the
house, describing the beauty: the 'unearthly sight, a scene of
heavenly splendour' from the drawing-room window over the
lake—'could you but have heard the thrushes and seen the
thousand thousand primroses under the trees!' However, 'we
will try to reconcile ourselves to your not coming down this
year'.[34] She was, in fact, never reconciled, and her whole reaction to Catherine's illness seemed hysterical. When she had a
letter, she was 'afraid to look into the terrible history of what
you had endured'. The Clarksons saddened her by selling Eusemere, but in 1805 Catherine was persuaded to stay for four
months in a rented house at Grasmere. Even Dorothy was forced
to agree 'you would not be quiet enough' in the crowded cottage, though 'nothing could be of more service to you than
spending the summer here'. Though Dorothy persistently
evaded the fact that Catherine did not like the Lake District by
regarding her as an invalid, Catherine, who lived to the age of
eighty-four, was lively company, 'cheerful—even merry and
the cause of mirth by our fireside'.[35] When she went home to
Bury, 'everything was almost the same—but yet it seemed very
different'.[36] It was five years before she felt free to accept the
Clarksons' repeated invitations to visit them in Suffolk. Yet she
continued to write regularly to her friend, with family news,
copies of William's unpublished poems, and transcripts of her
own journals, all faithfully preserved by this most discerning correspondent. When part of an original journal was lost, Catherine
could even supply it from her own copy.

The second shock of autumn 1803 concerned Coleridge. The
Scottish tour had been shadowed by the feeling that all was not
well between them. It was not merely anxiety over his health
which caused her to write on 29 August when after 'our parting',
having divided their common cash, the Wordsworths 'drove
heavily along . . . our thoughts were full of Coleridge'.[37] It has
been thought that the parting itself was precipitated by his
hearing Dorothy, at Loch Lomond, reciting appreciatively

William's poem *Ruth*; Coleridge wrote there, in his private notebook,

tho' the World praise me, I have no dear Heart that loves my Verses—I never hear them in snatches from a beloved Voice, fitted to some sweet occasion. . . .[38]

To friends he wrote, more than once: 'I soon found that I was a burthen on them.' He found William intolerant and spoilt by family adulation, and that this was the result of his 'living wholly among *Devotees*—having every the minutest Thing, almost his very Eating and Drinking, done for him by his Sister, or Wife'.[39] These are Coleridge's first suggestions that Dorothy is a harmful influence on her brother, recorded in his notebooks five years later, when he actually added remarks, which make this plain.

Equally, Dorothy's own unquestioning admiration of Coleridge began to be shaken. During his month's stay at Dove Cottage, from mid-December 1803 to mid-January 1804, she commented on the difficulty of nursing him, in a house with two babies, Johnny and his own Derwent, as an unasked guest while he was[40]

lame with the gout, stomach-sick, haunted by ugly dreams, screamed out in the night, durst not sleep etc etc . . . continually wanting coffee, broth, or something or other. . . .

She or Mary sat up by him each night, to wake him from terrifying nightmares.[41] 'You may think we were busy enough,' she added, not without an undertone of criticism.

Yet they still had amicable discussions, for instance on Beauty —'She does not think a smooth race horse beautiful but prefers a rough colt'—and what poems to repeat to 'a beloved Housemate'—Sara Hutchinson was also there at the time.[42] 'Dorothy thinks', doubtless with Coleridge's unfinished *Christabel* in mind, 'it would be better to wait till something was finished, that it could be repeated as a whole.'[43] When he had left the cottage, but was still in England, she was 'literally at work from morning till night', copying William's poems for Coleridge to take on his journey. By heroic efforts, about eight thousand lines of fairly illegible manuscript were fair-copied and recorded for the future on Coleridge's wish. 'Think what

7. 'returned to our native mountains, there to live': Town End, Grasmere

8 (a). 'the old and dearest spot of all': Dove Cottage

8 (b). 'are they setting up for fine Folks': Rydal Mount

they will be for me!' he urged. It is doubtful whether any of the women at Grasmere realized how, for Coleridge, any small incident in their lives could become an ever-enduring symbol. The incident at Gallow Hill when Mary (before her marriage to William) and Sara both petted him innocently in the evening firelight recurs, obsessively significant, in several later poems. Yet one feels their relief from the heavily charged atmosphere when, in April 1804, Coleridge sailed, in a convoy guarded by a British warship, for the naval port of Valletta in Malta, intending to proceed to the warmth of Sicily. By mid-April he had safely reached Gibraltar.

Dorothy's thoughts, however, were still abroad with the perils of war, this time transferred to her brother John. On 15 February 1804, the East India Fleet, returning from China, was attacked by a superior French squadron in the Malacca Straits, but succeeded in beating it off. Dorothy was 'anxious about my brother John—lest he should fall into the hands of the French'.[44] John, with the *Abergavenny*, reached England on 14 August, 'in great Spirits' after a successful voyage, to learn, within a few days, that William's first daughter, Dorothy's god-daughter and namesake, had been born on 16 August.[45] He was too busy, though, with plans for 'a better voyage' to come to Grasmere, where, as Dorothy wrote,[46] 'we have no nursemaid for the children nor have we had any nurse to take care of my Sister during her confinement; I was the attendant upon the little Baby and Sara Hutchinson came over to assist me'. Dorothy's health suffered, with loss of sleep, and eye-inflammation from fatigue;[47] William took her for a brief recuperative walking tour of Ennerdale and Wasdale:[48] 'what a lovely and wild road it is among high mountains'.

In spring 1804, during Mary's pregnancy, Dorothy both helped Sara and Tom Hutchinson to move into Park House, a new farm high in the hills near Penrith, and spring-cleaned Dove Cottage with the full North-Country ritual of scrubbing and painting. As she wrote, 'my Sister not being very strong, I was glad to take upon myself the charge.' The heavy work was the harder because old Molly Fisher's strength was failing and she could not help, though Dorothy was 'afraid of breaking her heart' by saying so. The problem solved itself when Aggie Fisher died and Molly was promoted to be her brother's house-

keeper and 'attendant upon his single cow'. A proud house-
holder, she entertained the Wordsworths to tea with 'piles of
toast', and in summer sent presents of her own butter and a
bottle of gooseberries.[49] Four years later Molly died and,
reversing the standard social custom, left her former mistress
the legacy of her best gown. Dorothy received it gratefully and
when it finally wore out, cut it into patches for a bed-quilt,
'old-fashioned work', stitching faithful Molly into the fabric of
her life. Another homely friend recalled herself, Peggy of Racedown, now married to a blacksmith, mother of many children
and very poor. Dorothy promised help and got Richard to send
her a pound note; in Dorset, where a man's wages were seven
shillings a week, this represented a substantial sum.[50]

Meanwhile, during the winter of 1804, John Wordsworth, in
London, received the ceremonial sword given by the Commit-
tee of Lloyds to the Captain of every Indiaman which had taken
part in the successful action in the Malaccas against the French.
On this next voyage, the *Abergavenny* was bound for Bengal as
well as China. She carried a cargo valued at £200,000, of which
£20,000 was John's own personal investment. The *Abergavenny*
sailed from Portsmouth on 1 February 1805. In the afternoon
of 5 February, the ship struck a rock in the Shambles, two miles
from Portland Bill, and sank in Weymouth Bay, with the loss
of over half of its complement, including John himself. Richard,
hearing the news in London, wrote to Grasmere. William and
Mary were out walking when the letter was brought from the
post by Sara Hutchinson, and Dorothy was the first to receive
the shock, with sorrow 'which', wrote William, for her, 'is and
will be bitter and poignant'.[51] William added that John 'was
worthy of his Sister, who is now weeping beside me'. She tried
to comfort herself with a survivor's report that when John's
ship was going down, he said to the Mate 'the will of God be
done' and 'was as calm as before.'

As William foresaw, the blow fell worst upon Dorothy, and
her brothers, grief-stricken themselves, did all they could to
ease its worst effects, which lasted all the rest of that year. As
late as 29 November, after illness, looking around the familiar
furniture of their home, and seeing how many simple objects
were associated with her dead brother, she had to confess, 'my
tears *will* flow'.[52] She had been specially shaken, earlier the

same month, by the news of Lord Nelson's death in the moment
of victory at Trafalgar, which also caused a burst of sympathetic
tears. Well might Jane Marshall, writing after John's death,
speak of 'her too *too* feeling Heart'.[53]

'What a fearful thing a windy night is now at our house!' exclaimed Dorothy. 'I am too often haunted with dreadful images
of Shipwrecks and the Sea when I am in bed and hear a stormy
wind.' This happened frequently in this December 1805, an
exceptional month of gales and storm-blown snow when 'the
chimney every now and then roars as if it were going to come
down upon us'.[54] She was, most of the time, alone with the
children Johnny and Dora, for Mary, pregnant once again,
was with Sara at Park House, where William had joined her.
In the winter evenings, while the children slept, Dorothy put
herself to compose poems for them.

> What way does the Wind come? What way does he go?
> He rides over the water, and over the snow,
> Through wood, and through vale; and o'er rocky height,
> Which the goat cannot climb, takes his sounding flight;
> He tosses about in every bare tree,
> As, if you look up, you plainly may see,
> But how he will come, and whither he goes,
> There's never a scholar in England knows. . . .

Nursing little Dora, she tried her hand at a lullaby, beginning:

> The days are cold, the nights are long,
> The northwind sings a doleful song;
> Then hush again upon my breast,
> All merry things are now at rest,
> Save thee, my pretty Love!

She deprecated these efforts '(muttering to myself as is my
Brother's custom) to express my feelings in verse'.[55] Yet they
are not unskilled, and catch the atmosphere of the home-keep-
ing Christmas at Grasmere when

I have been summoned into the kitchen to dance with Johnnie and
have danced till I am out of Breath. According to annual custom, our
Grasmere Fidler is going his rounds, and all the children of the
neighbouring houses are assembled in the kitchen to dance.[56]

On this thirty-fourth birthday, she could still dance in spite of
sorrows.

In July, Dorothy had a visit from Aunt Threlkeld of Halifax and her daughter Elizabeth; both were shocked by her prematurely aged appearance, 'so thin and old that they should not have known her—lost many of her teeth and her cheeks quite sunk that it has entirely altered her profile'.[57] She had never, of course, recovered the weight she lost in her severe illness at the end of her time at Racedown, and the recurrent 'fits of illness' at every crisis of the last three years. She carried lack of personal vanity to the point of self-neglect. 'My tooth broke today,' she wrote. 'They will soon be gone. Let that pass, I shall be beloved —I want no more.'[58] All the same, her appearance in summer 1805 must have been partly due to the continuing shock from John's death. Her brother Richard, so often prevented by his busy law practice from accepting Dorothy's fervent invitations to come and stay with her, made time to do so, also at the end of July. He too was disturbed at her lack of health, 'and fearing that I injured myself with walking, being at that time not very strong', he bought a pony for her, so that 'I who in my youth, could never ride half a dozen miles without great fatigue have ridden twenty six and was not the worse for it'.[59]

Richard had fitted her out with a side-saddle and bridle; William, for his part, could always borrow a neighbour's horse, and they took several expeditions together, to provide her with rest—the summer had been a busy one, with two young children and many visitors. On 6 November, they set off on a week's visit, to Ullswater and beyond, on this occasion sharing the pony. Something of the former Dorothy shows through the entries in the brief journal she kept. They visited Sara Hutchinson, housekeeping for her brother's farm at Park House in the hills above Penrith, and Dorothy renewed her acquaintance with Penrith itself, remembering her girlhood days when she had first known the Hutchinsons there, and shared early confidences with them.

Part of the time they stayed in Patterdale with their friends Captain and Mrs Charles Luff. Luff was an officer in a local volunteer company, the Patterdale Loyal Mountaineers, and it was while at the Luffs' house near Goldrill Bridge that Dorothy heard of Nelson's death. The Luffs mainly appear in Dorothy's letters as suffering ill-health, notably the Captain's attacks of gout, but they evidently lived in a beautiful place, near which

William, with his usual restless optimism, 'pitched upon the spot where he should like to build a house better than any other he had ever yet seen'. Nothing came of the plan to build on the spot, which he ultimately acquired, but the lovely and romantic little valley itself is transmuted immortally in the pages of Dorothy's journal.[60]

The vale looked as if it were filled with white light when the moon had climbed up to the middle of the sky; but long before we could see her face while all the eastern hills were in black shade, those on the opposite side were almost as bright as snow. Mrs. Luff's large white dog lay in the moonshine upon the round knoll under the old yew-tree, a beautiful and romantic image—the dark tree with its dark shadow, and the elegant creature as fair as a spirit.

The whole expedition with its return home over Kirkstone Pass, where her unmistakable voice is heard in the final page of her journal, seems to echo a return from grief to happiness.[61]

The torrents of Kirkstone had been swollen by the rains, and filled the mountain pass with their roaring, which added greatly to the solemnity of our walk. The stars in succession took their stations on the mountain tops.

The last sentence is as effectively poetic as anything William was now writing, in his effort to get the huge 'poem to Coleridge' (finally known as *The Prelude*) finished before the return of Coleridge himself from his prolonged sojourn by the Mediterranean.

Dorothy's thoughts on her thirty-fourth birthday, Christmas Day 1805, were resigned, even after the loss of John; 'these years have been the very happiest of my life . . . though till within this late time I never experienced a real affliction'.[62] There was still, however, one continual and increasing worry in the New Year, and indeed all through 1806. Where was Coleridge, what was he doing, and why did they hear so little from him? Dove Cottage was full of this anxiety, especially among the women there, for Sara Hutchinson had now come, in January 1806, to visit and live permanently, with Dorothy, William, Mary (whose child was expected in June), and the two children. Dorothy welcomed her help, though somewhat apprehensively looking toward the birth of the coming child.[63] 'We are crammed in our little nest edge full as you will suppose. . . .

we are a housefull now, what shall we be then? Every bed lodges two persons at present.' When Mary, after a painful confinement, gave birth to a second son, Thomas, on 15 June, the 'housefull' became virtually unmanageable. 'Our continuing here during another winter would be attended with so many serious inconveniences, especially to my Brother, who has no quiet corner in which to pursue his studies, no room but that where we all sit (to say nothing of the unwholesomeness of these low small rooms for such a number of persons . . .'[64] Dove Cottage had seemed small when William and Dorothy first entered it in 1799; as a home, it was now breaking down under the pressure of William's marriage and growing family.

In March and again in July 1806, Coleridge was said to be coming home; but 'every post day we trembled,' Dorothy wrote, 'when I was alone in bed at night I could not banish the most dreadful images—and Mary and Sara have suffered in the same way'.[65] Very few letters came; Coleridge said variously that he was too ill to write, or that all his 'long letters' had 'sunk to the bottom of the sea'.[66] In fact, facing a mass of problems of his own, he had virtually cut himself off from England for more than two years. His outward life had been surprisingly successful. In Malta, he had established himself as a useful member of the wartime Civil Service, unofficial private secretary and then Acting Public Secretary to the High Commissioner, Sir Alexander Ball. Physically, he improved, and on visits to Sicily, twice climbed Etna. Mentally and spiritually, he came to an agonizing reckoning with himself. He now faced the fact that drug-taking, although he could not stop it, was the root of his trouble, and that he needed entirely to reorganize his life. He decided to leave the Lake District, the Wordsworths, and his wife, to take on regular journalism and lecturing in London, and build a fresh existence.[67] When he arrived in Portsmouth in August 1806, after a terrible crossing in a neutral ship, he did not go north for two months, but remained in London, improving his *Ancient Mariner* and building up literary contacts. Appeals from Grasmere eventually brought him to an inn at Kendal on 26 October,[68] where Dorothy first saw him after two and a half years. 'We all went thither to him and never never did I feel such a shock as at first sight of him,' she wrote. 'We all felt exactly in the same way.' Even 'the divine expression of his countenance. Alas! I never saw it as it used to be.' His whole appearance had totally changed, as a picture painted in Italy before his homecoming shows, face fatter and coarse, hair cut short and going prematurely white at the temples. 'His fat-ness', wrote Dorothy, 'has quite changed him—it is more like the flesh of a person in a dropsy than one in health; his eyes are lost in it.' He was unwilling to talk personally, and kept chang-ing the subject. So the two, both in their thirty-fifth year, faced the travesties of their former hopeful selves, one a heavy, unhealthy mass of puffy flesh, the other haggard, almost tooth-less, her face darkened to a 'determinate gipsy tan', and her features deeply drawn by continual toil and hard living. It was a nightmare as haunting as any line in Coleridge's great poem.

12. *'I shall always date* Grasmere*'*

Some of the shock felt by both Dorothy and Coleridge was because, in the three-year gap with little news, they had developed obsessive and different needs. Each had too much of an overriding concern to pay full attention to the other. With Coleridge, this was his plan to separate from his wife. He felt he had at last come upon a solution to the problem, often expressed before in various forms, that he could not live without his children, and yet could not live with their mother. He had now reasoned himself into a determination 'to part absolutely and finally; Hartley and Derwent to be with me but to visit their Mother as they would do if at a publick school'.[1] Little Sara, who had so enchanted him before he left for Malta —'O I could drive myself mad about her'[2]—did not enter his present calculation; nor did her mother, who, he seemed to think, would be unreasonable not to agree. He was too absorbed in his own new plan to realize the Wordsworths might have one for their own lives.

For Dorothy, Mary, and William, the dominant fact was that Dove Cottage seemed too small for them to face another winter there, especially after the birth of Thomas. Luckily, a week before that event, in June 1806, a temporary solution had been provided. In Dorothy's words, 'We think of going into Leicestershire, Sir George B. having offered us their house for the winter.'[3] In fact, the whole Wordsworth family, including Sara Hutchinson, were actually bound for Leicestershire when they met Coleridge at the Kendal inn. Their providential benefactor and his wife had had a considerable influence on William and Dorothy during the three years which had just passed, and would continue to do so for years to come.

Sir George Beaumont, seventh Baronet, was the pattern of a Regency patron and connoisseur: urbane and assured, with an ironic wit, but a generous heart. 'His taste', as B. R. Haydon noted, 'was exquisite.' His family fortunes came from coal-pits scattered over their Leicestershire estate at Coleorton, which had been profitably worked since Tudor times, until even the trees of the original game park were lost to mining.[4] Beaumont was an amateur painter of moderate gifts and immense industry; more than two thousand of his pictures still survive. He had taken private drawing lessons at Eton and Oxford, where a German artist taught him to compose landscapes of natural simplicity, around a cottage or a ruined farm. Close study of nature for painting led this eighteenth-century character to appreciate the new Romanticism in literature. He was an early admirer of the *Lyrical Ballads*, and *Tintern Abbey* was his favourite poem.[5]

He met at a country ball, and married in 1778, a woman who shared his tastes. Margaret Beaumont cultivated Romantic sensibility; Coleridge later noticed how she trembled with enthusiasm and 'cannot keep the Tears in her eye',[6] while Haydon found her 'enchanting'. They spent their honeymoon in the Lake District at Keswick, and the next two decades founding a London Salon for artists and writers at their Grosvenor Square house. All this time they were assembling a distinguished art collection: Cozens drawings, Girtin watercolours, theatrical paintings by Zoffany, Rembrandt's *Jewish Merchant*, and—a gift from Lady Beaumont to her husband—Rubens's superb *Château de Steen*. This was the couple, now middle-aged but enthusiastic as ever, who in summer 1803 had decided to revisit the scene of early happiness and rented part of Greta Hall, Keswick, for a long painting holiday.

The Beaumonts were not at first pleased to find the other half of their house occupied by the talkative Coleridge, with his illnesses and inexhaustible catalogue of symptoms. Yet they were won over when 'Wordsworth not himself was his Theme', and were, as Coleridge reported, 'half-mad to see you'. Lady Beaumont read William's poems and announced, 'had you entered the room she believes she should have fallen at your feet'. Even Dorothy could feel satisfied with this fellow-devotee. 'She has a soul', wrote Coleridge, with his usual psychological flair, 'in point of quick enthusiastic Feeling, most like to Dorothy's—only not Dorothy's powers.'[7] Sir George, though they had not met, gave Wordsworth a plot of land, a picturesque rocky dell on Skiddaw, less than two miles from Greta, for the brother poets to meet more easily. The gift, like all his gifts, was gracefully made. There was no chance to offer thanks in person, for when William and Dorothy arrived at Keswick on 12 August 1803, they found the Beaumonts had just left; 'to be

so near and yet to miss you!' wrote Dorothy ruefully. 'We
arrived at the house scarcely more than two hours after you had
left it.'

This intimacy did not arise at once. When presents of game
or a home-brewed cask of stout arrived from Coleorton Hall,
Dorothy at first acknowledged them rather stiffly, 'with most
respectful Regards', as though conscious of the difference in
their stations.[8] Yet all that she learnt of Margaret Beaumont,
her quiet religious faith, family prayers at the Hall and school-
building in the village, drew them together in their letters.
Thomas Monkhouse found this society hostess 'a good creature
—sensible, though oddish'. Dorothy, who previously had
denounced the evil of riches, was disarmed to find the fashionable, aristocratic Beaumonts 'human-hearted creatures', as she
reported to Catherine Clarkson, 'whom it does one good to
know'.[9] With Catherine, Lady Beaumont became one of her
chief correspondents. A slow change in Dorothy's views grew
from friendship with Lady Beaumont, and both Wordsworths
took the first steps to the conservative principles of their later
years.

The correspondence soon grew affectionate. Dorothy 'longed
ardently' for them to meet. Her only fear was that she might
'seem inferior' and so disappoint. 'I have not those powers
which Coleridge thinks I have—I know it,' she protested. 'My
only merits are my devotedness to those I love and I hope a
charity towards all mankind.'[10] She felt it painful to love her
new friend, 'not having a bodily image of you in my mind'.[11]
The two were joint godmothers, Sara Hutchinson standing
proxy for Lady Beaumont, at the christening on 16 August
1804 of Mary's baby daughter, Dorothy: an old-fashioned name
'but my Brother will have it so'. The doings of the babies were
endlessly chronicled by one childless woman to another. It
seems to have been specially for the benefit of Lady Beaumont,[12]
'for one acquainted with the general features of the country',
that Dorothy wrote her lyric account of the Ullswater 'ramble'
in November 1805. To this unknown but sympathetic woman,
she opened her deepest feelings: their shared anxiety for the
absent Coleridge, 'no letters'; her grief at John's death, how
'fraternal affection . . . has been the building up of my being,
the light of my path';[13] even her earliest poignant regrets for

the mother she loved but could hardly remember. She was frank about the discomfort and overcrowding at Dove Cottage, especially in wet weather or in summer 1805 when 'Tourists', Walter Scott and wife, and Humphry Davy, came to stay. Johnny was her bedfellow in the room over the cold pantry, but where can the visitors have slept? A rough sketch made the problem plain.[14] They built a moss hut at the top of the orchard, modelled upon a summerhouse they had seen in Scotland, and here, with wrens twittering on the thatch, they spent whole days, lingering 'till the stars appear'. This, though, was only a summer expedient. Thus it was that Sir George's offer, coinciding with the birth of the third child, Thomas, seemed the answer to prayer.

What Sir George proposed for the winter of 1806-7 was that the entire family should stay as guests on his estate. 'You can hardly guess', wrote Dorothy, 'with what pleasure I look forward to the meeting.' The Wordsworth household arrived at Coleorton on 30 October, after what even Dorothy admitted was 'a troublesome journey', the elder children fretful and the baby 'very ill in his cough'. Wordsworth and Sara travelled separately by coach.[15] The Beaumonts were there to welcome them like old friends and install them in the home farm, since the Hall itself was being demolished and rebuilt by George Dance, architect of Newgate Gaol, with a domed hall and gallery to display their pictures. For the first time, Dorothy met her correspondent, Lady Beaumont, with her kindness as hostess, her charm of manner, her slight, engaging stammer; 'delightful, affectionate good people,' she reported to Catherine Clarkson.[16] The comfortable estate cottages showed equal care for their tenants. The farmhouse was roomy and warm; milk, butter, and poultry came from the farm, coal from Sir George's pits, the dairymaid cooked, and they hired a washerwoman. True, Johnny, when in a rage, would hit his sister with 'stool, chair, table, stick or even poker'; yet his aunt persuaded herself that 'he has a sweet temper'. For the next eight months, from November 1806 to June 1807, Dorothy and Mary enjoyed a rare leisure; 'nothing to do', she wrote exultantly, 'but read, write, walk and attend to the children'. All agreed she had never looked so well.[17] There was a plan, abortive, for William to visit London, but Dorothy had already sacrificed this pleasure for herself, 'because we cannot leave Mary . . . and I *have* seen London'.[18]

As at Alfoxton and Grasmere, the Wordsworths seemed odd-ities to the locals. The Scots gardener was suspicious of William's gardening, and an old labourer followed amazed as he tramped up and down the walks, composing aloud. The three women seemed 'marvellous creatures', ready to wade undaunted through the mud to fetch letters from Ashby de la Zouch post-office. Dorothy was transported with delight at the wide sky, the glori-ous sunsets, the rosy clouds, and at night the wild glow of coal-pit fires on the surrounding hills. The only anxiety was Coleridge who, in four letters, announced his determination to separate from his wife.[19] William, in reply, urged him to come to the farm for a visit.

Coleridge arrived at Coleorton with one of his sons, Hartley, on 21 December 1806. To Dorothy's relief, he seemed 'more like his old self'. All the same, her personal thoughts were sombre, although it was Christmastide. 'My Birthday,' she com-mented, 'but in my inner heart it is never a day of jollity',[20] though she drank a health to their absent hosts, the Beaumonts, in a bottle of port. The next few evenings up to the New Year, however, suggested a revival of the old magical days. So far, Coleridge had known only the first five books of William's 'poem to Coleridge', which he had taken, in their faithful copy by Dorothy, to Malta. Now, in his absence, William had com-pleted the whole poem of thirteen books, each containing some-thing like 500-800 lines. All these William now read by the Coleorton fireside to the little audience of Coleridge himself, Dorothy, Mary, and Sara. All were deeply moved, Coleridge to the extent of composing a poem himself, beginning

> O Friend! O Teacher! God's great gift to me

and ending with a description of

> That happy vision of beloved faces

that formed the audience, confirming his final line—

> And when I rose I found myself in prayer.

Yet, in spite of this happy and memorable interlude, every-thing concerning Coleridge had now an air of unreality. 'Cole-

ridge', Dorothy wrote on 16 February 1807, 'has determined to
make his home with us; but where?'[21] This was still unanswered
when they returned to Grasmere in June.

Such sober realism, and a new-found respect for the landed
gentry, exemplified by the Beaumonts, were not the only
changes in Dorothy's outlook as she lived through her thirties.
John's death had formed a watershed in his family's life. For
Dorothy it was the first step towards a total transformation of
her inner life. She seemed indifferent to conventional religion,
calling the Rector of Grasmere 'a worthy man . . . very good as
a Steward or farmer but totally unfit to preach or read prayers'.
Yet now, seeking comfort for John's death and later family
griefs, she turned increasingly to the Church of England. Slowly
at first, profoundly later, her view of politics, morality, and
religion changed from Romantic enthusiasm to the conventional
piety of her middle age. This change was first announced in
a letter to Catherine, as early as 1807. 'We are become regular
church goers . . . every Sunday.'[22] Here she shared an experience common in her times. In the distressed and dangerous
years of Napoleonic wars, and the depression that followed,
many felt that England, noisy with machine-breaking and
bread-rioting, was on the brink of revolution. Southey wrote in
1809 of the need 'to save this country from that revolution to
which it is so certainly approaching', and, a few years later,
Coleridge wrote of Jacobinism appealing 'to the brute passions
and physical force of the multitude (that is, to man as a mere
animal)'.[23] Wilberforce had already described the danger as
'rather a moral than a political malady . . . to the decline of
religion and morality our national difficulties must be chiefly
ascribed'.[24] To the middle classes, shocked and frightened by
radicalism, it seemed in Southey's words later that 'Religion is
the one thing needful . . . without it this world would be a mystery too dreadful to be borne'.[25]

A new seriousness appeared in the liveliest characters. Jane
Austen, having received *Pride and Prejudice* from her publishers
in 1813, announced to her sister 'a complete change of subject'
for *Mansfield Park*—'ordination'. The gentle, serious heroine of
her last novel, *Persuasion*, she felt 'almost too good for me'.
During a remission in her final illness, she wrote, 'the Providence of God has restored me—& may I be more fit to appear

before him when I *am* summoned'. Sydney Smith, robust and genial, preached of individual faith, 'the dawning thought of piety, the ascending prayer, the descending grace', and wrote in a private notebook that 'The tomb speaks more than a thousand homilies'. In the Church of England 'seriousness' was especially strong among Evangelicals, where 'Every hour and every shilling for God' became the rule, with a belief in the essential sinfulness of human nature, which demanded stern government in family and nation alike.

In Dorothy the seriousness of the nineteenth century was marked. She had early links with Evangelicalism from girlhood memories of piety at Forncett. She followed one of its great causes, anti-slavery, through the Clarksons, and in 1807 read Thomas Clarkson's two-volume history of the movement in manuscript. This was indeed the book she found as gripping as her old favourite *Clarissa*,[26] a startling change of taste which few readers could claim to share. With leisure to read at Coleorton, her first recorded choice was Fox's *Book of Martyrs*. Christian education was another Evangelical good cause, and Sunday Schools were provided by manufacturers, to tame the children of an older, wilder England into a docile workforce. Dorothy already heard small Dorothy's prayers 'in her night coat' and Johnny's catechism, but the arrival in 1811 at Grasmere of a new curate, 'very earnest in his attentions to Sunday Schools', encouraged her to take them for regular instruction from nine o'clock until church time.[27] This curate, William Johnson, became a family friend; unlike his tipsy predecessor, he proved 'so much in earnest . . . that I think we should go, even if we had not the children, who seem to make it a duty to us'.[28] With each child's birth, this sense of duty increased.

Like many democrats, with political hopes abandoned, Dorothy turned her enthusiasm to traditional relationships and customs, the affections, sentiment. The growth of her serious religious observance coincided with the growth of William and Mary's family. Children impose their own discipline on adults. Late-night walks and talks were forgotten when Dorothy was up at seven each morning to dress and feed her charges.[29] Home duties were sacred, taking precedence over writing, or even reading, while a sock remained undarned; besides, 'any new book in our neighbourhood passes from house to house and it is difficult to come at it in any reasonable time'. It was January 1806 when the ever-obliging Charles Lamb forwarded from London the first books she ever ventured to buy for herself, a box valued at six guineas. Even then, she added, 'we get nothing read'. After the birth of a third child that summer, when Mary suffered considerably more than ever before and was exhausted, Dorothy recorded, 'I have read nothing for weeks.'[30]

Dorothy's new strictness in sexual morality appears in her open criticism of Coleridge, who, after his Coleorton visit, avoided his wife by taking refuge with tolerant Thomas Poole, at the safe distance of Nether Stowey. He dreaded failing his friends' hopes and suffering their judgements, with '*pain* as from blows . . . I have not only not answered my letters,' he wrote, 'God help me, I have been afraid even to open them.' Dorothy noted dismissively in the autumn of 1807, 'Coleridge has never written to us and we have given over writing to him, for what is the use of it? We believe he has not opened one of our letters.'[31] She had already written of his 'drinking Brandy', and she now set down in writing for the first time a more serious criticism: 'I have no doubt that he continues the practice of taking Opiates as much as ever.'[32] The Wordsworths faced what they had dreaded, another winter in the cottage, never quiet until the children were in bed, and by January 1808 the stoical Mary at thirty-eight was pregnant for the fourth time in six years. Coleridge, with his marital problems, was an unwelcome thought. Dorothy confided to Mrs Clarkson her 'fear of having our domestic quiet disturbed if he should wish to come to us'.[33] Nor could she believe in the reality of his passion for Sara, judging, with unconscious contempt, that 'his fancies will die away of themselves'. Coleridge was stung into noting secretly, on 12 May 1808, Wordsworth's 'High Self opinion pampered in a hot bed' of his wife's and sister's adulation.[34] The end was still two years off, but slowly, inexorably, the friendship that had filled three lives with inspiration eroded from within.

Meanwhile, a new character appeared upon the crowded scene. On 4 November 1807 the wicket-gate opened, and a young man, pale and trembling with expectation, came up the path like one in a dream. Thomas De Quincey was, she wrote, so 'diminutive' that Dorothy had the rare experience of looking

down at him from her own five feet, so modest, shy, and rever-
ent towards William that he could hardly speak. He had first
read the *Lyrical Ballads* as a frustrated schoolboy, who fled from
the domination of his rigidly evangelical mother and guardians
and had narrowly escaped death from cold and hunger in the
streets of London. De Quincey was now at Oxford, a brilliant,
reclusive, wayward scholar, who had already discovered
opium, 'the secret of happiness'. He had written to Wordsworth
in 1802 to 'bend the knee' before 'genius so wild and so magnificent', a letter which Dorothy copied; he had even come
secretly to look at the cottage, but lost courage and retreated
'like a guilty thing'. He had now met Coleridge at a dinner
party in Somerset and offered to escort Mrs Coleridge and the
children home to the North, as a means of calling on Wordsworth. To Dorothy this 'pure and innocent' admirer gave extraordinary pleasure as an 'instance of the power of my Brother's
poems, over a lonely and contemplative mind, unwarped by
any established laws of taste'.[35]

This new disciple came in a time of need. Wordsworth's two-
volume edition of poems in 1807 was in the course of receiving
reviews, only one out of thirteen favourable; his critical reputation was at its lowest ebb. The *Critical Review* recommended
him to spend 'more time in his library'. The *Edinburgh Review*,
seen by chance at Penrith, appeared to 'revile the poems'.[36]
William was deeply discouraged, publishing nothing for the
next seven years, and Dorothy was defensive on his behalf.
Though forced to admit that his library was 'in fact little more
than a chance collection of odd books', she defiantly maintained
the *Edinburgh* notice was 'very silly and *ignorant*'. The hero worship of De Quincey, an excellent scholar, was balm to her hurt
pride.

On his part, he remembered the cottage idyll with delight:
the dark, oak-panelled kitchen, the diamond-paned windows
overhung with roses and jasmine, the 'plain household simplicity' of manner with which the poet's wife and sister made him
welcome. It soon grew dark; the family gathered with 'leisure
and conversation' round the table for a homely north-country
high tea, apparently their main meal of the day. He slept in the
best bedroom, but was woken by a little voice reciting the
Creed and found he was sharing it with Johnny. He went

downstairs, surprised to find Dorothy boiling a breakfast kettle
upon the open fire, for 'I . . . had never seen so humble a
ménage'. On the third morning they set off for a circuit round
Ullswater, by common farmer's cart, with a strong young
woman as driver. Everywhere they went, Dorothy was greeted
with smiles, and exchanged flying conversation with travellers
on the road. He found her 'the very wildest (in the sense of the
most natural) person I have ever known'.

Wordsworth later found these published details 'highly offensive', even 'infamous', and their author 'a pest in society'.
Certainly De Quincey's description of Dorothy, in body and
mind, is remarkably frank. Responsive himself to women, he
found her by contrast with Mary's sunny radiance, unfeminine
and physically unattractive. He dismissed at once any suggestion
of a love relationship with Coleridge, as he later dismissed
a sexual bond with William, since 'the young lady . . . had no
personal charms' and 'did not cultivate the graces'. She lacked
all feminine accomplishments or appeal; 'the least painful
impression was that of unsexual awkwardness'. Her sharp,
hurried movements and stooping walk 'gave her ungraceful
even unsexual character to her appearance'. Her tanned face,
with its associations of field work and low living, was then considered a disfigurement; most receipt books of the period contain a remedy, lemon-juice and spirit of wine, for bleaching
brown complexions. Her eyes, 'wild and startling, and hurried
in their motion', even her occasional stammer, suggested nervous tension and 'self baffling of her feelings'.

Yet of all the Wordsworth family, Dorothy had been his closest friend, and no one ever appreciated better the depths of her
intellect and character. He felt 'some subtle fire of impassioned
intellect' burning within her. Moreover he understood how the
sense of beauty, the sympathy swift as the pulses of light, and
the innate charity of 'the lady who paced by his side continually
through sylvan and mountain tracks', had humanized Wordsworth's austere genius, so that all 'worshippers through every
age of this great poet are become equally her debtors'. Seeing
the love of this ardent and fiery creature for her favourite
books, De Quincey, whose vast library overflowed the lodgings
of a lifetime, was astonished that she had so little systematic
education and seemed satisfied with the three-hundred-odd

battered volumes on William's shelves. Dorothy may have felt this implied criticism, for in July 1807, when De Quincey was in London, she asked him to buy for her brother history and classical translations *'very cheap'*.[37] Education aside, Dorothy struck De Quincey by 'the impress of originality upon all she uttered', her native freshness of intellect, the brilliant or be-witching effects of her casual speech or writings. He believed that, if instead of devoting herself to nephews and nieces she had become 'in good earnest, a writer', Dorothy had the talent to succeed and would have been spared much suffering in her later years. It is impossible to judge if he was right, since no one else, least of all Dorothy herself, appeared to consider the possibility.

Her gifts were fully displayed after a village tragedy in March 1808. One remedy for the Wordsworths' enforced retreat to the Grasmere cottage all through the winter of 1807-8 had been to hire a fourteen-year-old local girl, Sally Green, to help with the youngest child, Thomas. Johnny and Dora went daily to school, and Sally took her charge on outdoor expeditions such as nutting.[38] Late in March 1808, an appalling tragedy struck Sally Green's family. Her parents were killed in a fall from a mountain precipice, leaving Sally and six younger children, one an infant still at the breast. Dorothy helped to organize relief in money and aid among the neighbours, her own contri-bution being to retain Sally, when she had meant to part with her at Whitsun, 'not as a servant' but to train her for a domestic job.[39] Dorothy also wrote 'a minute narrative' of the tragedy, for distribution among the Wordsworths' friends, especially the Clarksons, who begged her to publish it: but Dorothy would not be drawn to let this 'simple and fervid memoir', as De Quincey called it, reach a larger public, though a short quota-tion shows its quality.

The eldest daughter had been unable to follow with the rest of the mourners, and we had led her back to the house before she got through the first field. The second fainted by the grave-side; and their brother stood like a Statue of Despair silent and motionless.[40]

In point of fact, it was not published until 1936. With its account of the small children waiting for news of their missing parents, of the discovery of the bodies, and of the funeral, it is intensely moving. Yet it does contain a certain new element in her writing. She seems, in part, to be arguing a case that the very poverty and simplicity of the Green family somehow constituted an act of virtue: even hard outdoor work is said to produce virtuous feelings.

I need not remind you how much more such a situation as I have described is favourable to innocent and virtuous feelings and habits than that of those cottagers who live in solitude and poverty without any out-of-doors employment.

There is a complete contrast in tone here with, say, the passage in the Grasmere Journal, eight years before, on the pauper funeral of Susan Shaelock[41]—'I thought she was going to a quiet spot and I could not help weeping very much'.

Dorothy's personal charity was to maintain the orphaned Sally Green under her own roof. An overdue change in the Wordsworths' lives made this possible, for in May 1808 they rented a larger house. Their choice of dwelling-place had its own irony. There was nothing Dorothy disliked more than the large new villas built in the country by the gentry or rich manufacturers. Her very first entry in the Grasmere journal on 14 May 1800 records the 'many sweet views up to Rydale head when I could juggle away the fine houses'. The new-style mansions emphasized 'prospect', a picturesque view and an imposing façade, with tall windows to light the great windy spaces within; the owner of one such house put on his hat to go into the next room. By November 1805, the fashion had reached even remote Grasmere, with 'Mr. Crump's newly erected large mansion, staring over the church steeple',[42] where its white stucco frontage glared against the hillside.* William called it a 'temple of abomination' to enshrine Mr Crump, a rich Liverpool attorney, who was said to have spent fifteen hundred pounds on building it. Dorothy sighed to Catherine Clarkson, 'I have often said I would cease to grieve about these things, but now I cannot help it.'[43]

Nothing could have been more alien to the Wordsworths' tastes and habits than this new mansion, Allan Bank; yet a series of circumstances drove them to become the first tenants. Coleridge needed to live with them, and to have his two boys to visit

* The house is shown on the hill left of the church in Plate 6.

on their weekend leave from school. Sara Hutchinson was threatened, they thought, by tuberculosis, and her younger sister Joanna by her own mysterious nervous condition; they must be offered a home if necessary. Most urgently, Mary expected her fourth child. Allan Bank would house thirteen people, indeed fifteen at weekends, when Hartley and Derwent Coleridge came. From stress and anxiety at the prospect, Dorothy had one of her characteristic attacks, 'a violent inflammation in my Bowels', and complained that her head felt 'muddy'. Yet by 1808 the decision to move to Allan Bank had been forced upon them.

The prospective move caused one of the rare surviving letters from Dorothy to William. He had gone to London in February 1808 when Coleridge was due to lecture at the Royal Institution, having heard that 'our poor friend' was so ill 'he *cannot* live many months'. He had also taken his new poem *The White Doe of Rylstone*, hoping to publish it for a hundred pounds, but once in London had doubts of its success and had given up the idea. Dorothy's response to this news was strictly practical. Publish, she urged him, and defy the critics—'what matter if you get your hundred guineas into your pocket?' William's resources, the Calvert legacy, the Lowther settlement, the meagre sales of his poems, had been swallowed up by the expenses of his ever-growing family—'and without money what *can* we do? New House! new furniture! such a large family! two servants and little Sally! we *cannot* go on so another half-year.' In the large house, if they had fewer servants, she and Mary must 'work the flesh *off our poor bones*'.[44] William could not bring himself to publish the poem until 1815, and after the move Dorothy's anxiety about money assumed the proportions of a nightmare. 'We have far outrun our income,' she wrote, begging Richard for a statement of accounts at New Year 1810, with endless schemes for economy. They must move where coals were cheaper, 'give over drinking tea', settle where there was a good Grammar School for the boys.[45] This last plan might have saved many future anxieties, but of course they could not bring themselves to leave the beloved vale. Even Coleridge, whelmed in his own troubles, was grieved at Dorothy's 'exceeding anxiety about pecuniary matters'. Meanwhile, she thankfully wore Mrs Clarkson's cast-off clothes.[46]

At the end of May 1808 the Wordsworths moved across the village into Allan Bank. Sara was still an invalid, Mary had fallen over some planks and sprained her right arm, and, as Dorothy wrote with unusual frankness to Catherine Clarkson, they hoped to get '*William out of the way*',[47] since his help was not worth cooking his meals. She was saved by a timely visit from the sailor brother who had escorted Mary on her first visit to Racedown so many years ago. Henry Hutchinson had served in a slave ship, where his shipmates called him 'Mr Wilberforce' for his kindness to the unhappy cargo. To Dorothy's delight this self-reliant mariner could sew, cook, wash dishes, put up beds, or lay carpets, making everything shipshape. For a fortnight they worked together body and soul, and Dorothy grew to love Harry because his shy manner and helpfulness 'reminded me of my dear brother John'. By 5 June they were installed. 'We live at Allan Bank but I shall always date *Grasmere*,' she headed a letter. She enjoyed the luxury of a bedroom to herself, with a view over the mountains to the east. Yet she was drawn back to the old home, where the laburnums were in all the freshness of their beauty: 'the dear cottage! I will not talk of it . . . all within how desolate!'[48] Inevitably, after so much stress, the usual attack of violent purging and vomiting followed. Yet she was determinedly cheerful; 'next month we expect the next Bairn,' she wrote. 'God grant it may be like the rest';[49] and on 6 September 1808, Catharine, Mrs Clarkson's godchild, was safely born.

By year's end Allan Bank seemed almost as crowded as Dove Cottage had been. In the autumn Coleridge had arrived at their urging to launch 'a literary, moral and political weekly paper', *The Friend*, which he would dictate to Sara Hutchinson and print at Penrith, walking fifteen miles over the fells to deliver copy. A prospectus for this difficult project attracted nearly four hundred subscribers, some distinguished, and he was full of hope. At the same time, he was attempting to break his opium addiction, with agonizing withdrawal symptoms, which vanished whenever he relapsed and took his habitual dose. Tragically, when he most needed friendship, these alternations were incomprehensible to others and increasingly annoying to the overworked Dorothy. She wrote, 'he has often appeared to be dying and has all at once recovered health and spirits'.[50] she did not believe in the future of *The Friend*, because of Coleridge's 'utter want of power to govern his mind, either its wishes or its efforts'.[51] His marital affairs, she felt, exposed the Wordsworths to gossip. Poor Mrs Coleridge and their little Sara arrived to stay more than a week 'upon friendly terms' although their separation was deeply humiliating to her. In November De Quincey joined the household in an equally characteristic manner. At Oxford he had revised for his finals eighteen hours a day, never going to bed. He had done brilliantly in Latin Orals on a Saturday, was due to take Greek on the Monday, but on Sunday panicked and fled to London; he had related all this in letters to the sympathetic Dorothy. In October, she invited this diminutive refugee from life; 'whom you would love dearly, as I am sure I do', to take shelter with them. He was to help William, then 'engaged in a work which occupies all his thoughts'. Dorothy provided for this gathering, supervising 'the *Cook* (as I have rather aristocratically called her)', and a 'muddling' housemaid, while fulfilling an old ambition to keep a cow, which had to be milked two fields' lengths away from the house, but was 'quite a treasure'. So, as she said drily to Jane, 'idleness has nothing to do with my putting off to write to you'.[52] 'Poor Dorothy', wrote Sara Hutchinson, matter-of-factly, 'is kept trotting the whole day through'.[53]

This large household revolved round two literary projects: Coleridge's *Friend*, which eventually, with Sara's aid, reached twenty-seven issues, and Wordsworth's pamphlet attacking *The Convention of Cintra*. This truce of August 1808, offering generous terms to the French in the Peninsular War, outraged his anti-Napoleonic fervour. Writing far into the night on his lonely Lakeland hillside, he appealed to the feeling of the people against the weakness of their leaders. 'Every human being in these islands', he maintained, would rather have 'perished to a man' than agree to such terms; 'Unconditional surrender' was the demand of the nation.[54] By day he dictated to Mary or Dorothy, as Coleridge dictated to Sara, across smoke-filled rooms.

For, with the coming of winter, Allan Bank revealed its secret weapon. The house stood 'on a hill, exposed to all *hails* and *blasts*, and the cold seemed to cut me through and through'. By early December 1808 it was so cold that they had to go to bed in the middle of the day with the three-month-old baby, to keep it warm. The chimneys, unlike those of traditional cottages, were small, to enhance the grand façade, and did not draw. When they lit fires, clouds of smoke billowed out into the rooms. The women wanted to leave, 'but the Men will not hear of it. William is trying for a Cure', as Sara Hutchinson wrote caustically: not merely a chimney doctor, but 'a Scientific Chimney Doctor'.[55] 'Have you cured it?' enquired Charles Lamb. 'It is hard to cure anything of smoking.' But this was far from a joke: the housework was doubled and the new maids miserable, chairs, carpets, panelling, the clean dishes in the pantry, all were covered in soot. Dorothy and Mary were driven to 'scouring' in the icy evenings when the children were in bed.[56] Desperation proved stimulating; with smarting eyes, Dorothy wrote to Catherine Clarkson a bravura description of everyday life at Allan Bank: 'a smoky house—wet cellars—and workmen by the half dozen, making attempts, hitherto unsuccessful, to remedy these evils . . . this house is, at present, literally not habitable, and there is no other in the vale. . . . There was one stormy day in which we could have no fire but in my Brother's Study and that chimney smoked so much that . . . I, with a candle in my hand, stumbled over a chair, unable to see it. . . . We cooked in the Study, and even heated water there to wash dishes, for the Boiler in the Back-Kitchen could not be heated, much less the kitchen fire endured; and in fact . . . We have been for more than a week together at different times without a kitchen fire.'[57] Nevertheless, she added, they were all in good health and spirits.

Through all the chaos, William's 'first and last thoughts' remained safely in Spain and Portugal. By 20 February 1809 De Quincey was sent to London to see the pamphlet through the press and correct the proofs. By 11 March he had received seven letters of corrections from the anxious author, and on 26 March, four letters in one day. Dorothy wrote gratefully at the beginning of April, 'You have indeed been a Treasure to us while you have been in London, having spared my Brother so much anxiety and care—we are very grateful for your kindness'.[58] De Quincey continued to show 'the utmost reverence' for William. He was so well trusted as a faithful disciple that

Dorothy would even admit him to family jokes: the headaches that William 'in his gloomy way calls apoplectic', and his fears that the pamphlet might lay him open to the charge of libel, 'as his mind clings to the gloomy, Newgate is his favourite theme'.[59] Throughout the spring and summer of 1809, she kept up a series of cheerful, friendly letters. The pamphlet was published in June 1809. The controversy had already died down and Dorothy wrote loyally to Catherine Clarkson, 'All the judicious seem to admire it. Many are astonished with the wisdom of it—but nobody buys!!'[60] In the end the remaindered copies were sold as waste paper and used for the lining of trunks. One effect it did have, however, was to fuel Dorothy's hatred of Napoleon and the French, with the secret exception of Caroline Wordsworth and her mother.

The summer of 1809 reduced the smoke menace, but produced instead a 'houseful of company', a succession of enthusiastic Lakers who kept Dorothy in a bustle. At the same time, she had another domestic project, nearer to her heart. William had taken a further six years' lease of the Town End Cottage, which he sublet to De Quincey,[61] while Dorothy undertook the work of repairs and furnishing with houseproud love. Each letter carried a report: 'your snowdrops are in full blossom', the carpet was pretty, but the bed-curtains unsatisfactory, the walls were whitewashed and bookshelves put up by the village carpenter. De Quincey insisted upon mahogany, though 'so very dear', for the shelves, and she bought mahogany furniture to match.[62] Silver spoons must come from London. In August she spent twelve days in Kendal buying hardware. Thirty chests of books would arrive with the new tenant, and De Quincey's library 'will be a solid advantage to my Brother,' wrote Dorothy as a matter of course. It was as though she herself were returning 'to linger about as though we were still at home there'. On a spring day she 'sat half an hour musing by myself in the moss hut. . . . The little Birds too, our old companions, I could have half fancied, were glad we were come back again, for it seemed I had never before seen them so joyous on the branches of the naked apple trees.'[63] De Quincey would preserve their coverts; she could not imagine this tiny, timid person might prove to have a will of his own. In October 1809 he moved in, and celebrated with fireworks for all the children of the vale. Dorothy's happiness was complete. As she told Jane, he was 'one of our own Family, so we have now almost a home still, at the old and dearest spot of all'.[64] Where her 'Family' was concerned, she was still a romantic optimist.

13. 'Those innocent children'

If De Quincey seemed now to Dorothy 'one of our own Family', it was becoming clear that Coleridge, in her mind, was already forfeiting that title. As the year 1810 wore on, Mary, at Allan Bank, approached her fifth confinement. Dorothy's remarks about Coleridge, on whose behalf, partly, the large house had been taken on a short lease, show impatience, disillusion, and even annoyance. When he spoke, contrary to former protestations, of going back to live with his wife at Keswick, Dorothy jumped at the chance. 'I hope he will choose the time of Mary's confinement.' Sara Hutchinson, worn out by him, had gone to stay with relatives, and 'we are all glad she is gone' for her own sake. 'He harassed and agitated her mind continually.' She was frank about Coleridge's behaviour. 'We have no hope of him . . . you feel perpetually new hollowness and emptiness. . . . Burn this letter I entreat you,' Dorothy continued to Mrs Clarkson. 'I am loth to say it, but it is the truth.' She did not exaggerate Coleridge's inconsiderateness.

He lies in bed, always till after 12 o'clock, sometimes much later. . . . He never leaves his own parlour except at dinner or tea and sometimes supper, and then he always seems impatient to get back to his solitude—he goes the moment his food is swallowed. Sometimes he does not speak a word.

She concluded that his love for Sara 'is no more than a fanciful dream. . . . He likes to have her about him as his own, but when she stood in the way of other gratifications it was all over.'[1] To her evident relief, Coleridge left for a five-month stay at Keswick early in May 1810. They did not meet again for a whole decade.

This separation had its origins in a devastating misunderstanding between William and Coleridge. Its causes, though complex, can be briefly told.[2] By this October 1810, Coleridge had resolved to try and conquer the opium habit by placing himself 'under the constant eye, of some medical man'. Basil Montagu and his new wife, who were visiting Allan Bank, invited him to stay with them in their London house, and consult their friend and physician, Anthony Carlisle. Wordsworth, according to Dorothy's later account of the matter, 'used many arguments' to Montagu himself against this plan; but Montagu insisted, Coleridge 'acquiesced', and the party set off, leaving Allan Bank on 20 October for London. William was clearly uneasy when the chaise left, and Coleridge, perhaps to reassure him, said he felt 'very well in his head'. William made the rejoinder, none too tactful, but probably a piece of rough if embarrassed humour, that at the autopsy on Schiller, 'his entrails were eaten up while his brain was sound'. Apparently unnoticed at the time, this remark became the corner-stone of a set of grievances. For, on arrival in London on 28 October, Montagu repeated all Wordsworth's 'many arguments' to Coleridge himself, on the Godwinian principle of telling truths, however unpleasant.

No exact record exists of what Montagu said to Coleridge, but he is supposed to have quoted Wordsworth in the most damaging way, beginning by saying, 'Wordsworth *has commissioned* me to tell you, first, that he has no hope of you.' In detail, Wordsworth was alleged to have spoken of Coleridge as being 'an absolute nuisance', that he was 'in the habit of running into debt at little Pothouses for Gin'. Likewise, William was reported to have spoken of Coleridge as a '*rotten drunkard*' who was 'rotting out his entrails by intemperance'. The last remark was obviously a garbled version of William's tactless parting words about Schiller. In any case, Dorothy knew virtually nothing of this for several months when she heard, not from Coleridge—she was used to not hearing from him—but from Mrs Coleridge that her husband 'writes as one who had been cruelly injured'[3] by William. Her reaction was to spring automatically to William's defence: 'at first when I read all this my soul burned with indignation that William should thus (by implication) be charged'. She continued to defend William, and wrote off Coleridge as she did any critic of her brother. The two aggrieved poets were eventually brought to outward reconciliation in 1812, by the efforts of friends; but for Dorothy there remained no feeling for Coleridge except perhaps pity. 'Poor creature! unhappy as he makes others, how much more unhappy is he himself!' When reconciliation was partly effected, she sent her love, but wrote significantly of 'the depth of the affection I had had for him'[4].

If Dorothy put her affection for Coleridge in the past tense, it was partly because her mind was fully engaged elsewhere. She was always devoted to her family, but for at least a dozen years after her nephew John's birth, apart from a few holidays, the Wordsworth children dominated her life. She realized this, apologizing to correspondents for too much nursery 'prattle', yet they still invaded her letters. 'No person', she wrote in self-defence to Lady Beaumont in 1808, 'can be more seriously convinced of the bad effects of over-indulgence . . . and though I am far from thinking that I entirely avoid the fault, yet I hope I do not grievously err.'[5] Mary, with remarkable generosity, understood this overpowering love, and just as she had shared their father was prepared to share 'These dear children who now seem to complete our being'.[6]

The children gained more than the adults from the move to Allan Bank, where they could run wild on the hillside without fear of passing carts and carriages. 'They are very happy in the Liberty and freedom of this place': John building houses of stones and slates, with Dorothy as humble assistant and Thomas as destroyer. Dorothy suspected that Mrs Coleridge distrusted the rough wildness of 'our rustic brood', and did not want her ladylike little daughter to visit them; 'there is nothing about her of the natural wildness of a child'.[7] At moments, she allowed herself to denigrate the Coleridge children, calling Sara a little vixen or Derwent 'cowardly and effeminate and indolent'.[8] This resentment, so unlike her affectionate self, suggests a loss of balance in her obsessive love for her own family.

It was not easy to maintain equal justice between the Wordsworth children. Thomas presented no problem; a sweet-tempered, affectionate child, he was 'put into Trousers' and played happily with pots and pans in the back kitchen, earning the nickname 'Potiphar'. 'Everybody loves him.' Dorothy grew pretty, small and feminine; the servants noticed her 'pretty ways' and she entertained the household with lively conversation. Before the move she and John had already begun to go daily to the village school in Grasmere churchyard, and here their aunt's difficulties began. 'The Master is very fond of D. she learns so quickly as he says,' Dorothy explained to Catherine Clarkson, but John 'is less in favour with the Master', because he 'keeps his attainments to himself', while his sister was 'proud and not unwilling to display what she can do'.[9] The fact was that John had a difficulty in learning to read which no one could then understand and which amounted over the years to a serious disability. By summer 1808 small Dorothy, in contrast, continued to be 'very pretty, very kittenish, very quick, very clever', a favourite with her father and inclined to 'wayward-ness'. 'I fear we shall have great difficulty in subduing her,' wrote her aunt. There were battles, 'imprisonments', and Dorothy senior was 'most earnest in persuading William to let her go' to boarding-school, both to improve her manners and make her a 'useful Girl in the Family'. It is strange to hear this echo of disapproving old Mrs Cookson during her own rebellious youth. So in April 1809, while John was still struggling to 'read without spelling', Dorothy, aged four and a half and proud to travel alone, was put on the coach for Miss Weir's boarding-school at Appleby, the far side of Penrith.[10] Her aunt missed her lively presence and clear voice about the house, but persuaded herself the little girl was 'no doubt as happy as the day is long'. Her father on the other hand, visiting for an hour, found his small daughter 'sorry to part from me, crying bitterly and clinging round me'.[11] It is extraordinary that Dorothy, who had suffered so much from uprooting in her own childhood, could have brought herself to approve this exile.

Meanwhile she faithfully attempted to teach John herself, but by 1810 was forced to admit, 'After wearying out my stomach . . . I find that we have not advanced the smallest point.'[12] 'Of all Trades', wrote Sara Hutchinson, hearing her struggles, 'keep me from being a Schoolmistress!' William could not hide his bitter disappointment. John was 'the slowest child almost I ever knew', he wrote when his son was eleven. 'Incredible pains have been taken with him, but he is to this day a deplorably bad reader of *English* even.' Sara recorded the fear poor John had of his father.[13] For a time Dorothy had hopes that De Quincey, with his excellent scholarship and gentle manner, might solve the problem. He loved all the children, giving them colour pictures for their rooms—a Giant's Castle and a Magician's Temple—with a toy carriage to pull.[14] Transparently, she tried to engage his interest in John, writing how the boy valued his letters from London, prayed for him each night, 'longs for your return' and 'seems to desire to learn'.[15] She was satisfied that 'His conversation has been of very great use to John', just as his library was to John's father. Yet De Quincey's special love and interest was given to an unexpected member of the family.

Mary's fourth child Catharine was described by her aunt as 'the only funny child in the family'. She was uncommonly good-tempered and cheerful but 'very plain'. At one year she did not talk or walk, but spun round upon the floor with her little bald pate thrust forward. At a year and a half, Dorothy wrote to Lady Beaumont, Catharine had 'something peculiar in the cast of her face which probably adds to the comic effect of her looks and gestures'.[16] Wordsworth called her 'his little Chinese maiden'. At two years old she still did not talk and there was perhaps some warning that she might be, as it proved, the most vulnerable child of the group. De Quincey adored this little creature, who shared something of his own changeling quality, and she returned his love, searching for him in every room when he was away and calling him by a single word of her own invention.[17] De Quincey made them all promise that he should be 'sole Tutor', not to Johnny but to Catharine. This cherished plan was not to be. On an April morning of 1810, Dorothy went into the kitchen to find Catharine vomiting violently. In a few minutes she went into convulsions of the whole body and remained unconscious for seven hours, of 'unspeakable anxiety' to the adults. The day passed in the drastic remedies of the period, repeated purging and lancing of the gums, and at night the exhausted child seemed to sleep; 'but what was our grief next morning on discovering that she had lost the use of her right hand!' In fact the whole right side of her body was paralysed and the apothecary's remedy of scrubbing with mustard had no effect.[18] Catharine had possibly suffered the toxic encephalitis of childhood, but Dorothy, distracted with fear and searching for a cause, scolded simple young Sally Green 'very severely' for allowing her charge to eat raw carrot, of which the elder children were making 'bullets'. This was the 'little Sally' who two years earlier, at her parents' death, the Wordsworths had agreed to 'keep . . . not as a servant,' but as one of their household. Now, in a tumult of distress, they dismissed her as 'utterly unfit', sending her to a relative who was fortunately willing to

take her in. This decision, so unlike their better selves, shows
their intense, and, for the women, almost total absorption in
the children. Mary 'rubbed' Catharine faithfully every day for
an hour, and attempted, uselessly, to 'confine' her good arm,
hoping she would use the other. She learnt to 'move though
very lamely from one end of the sofa to the other'. At first,
Dorothy persuaded herself that when her teeth came through
Catharine would recover completely; she was so touching with
her 'sweet yet cunning smile', and 'would be very entertaining
if she could talk'. Yet after eighteen months she was still lame
and Dorothy was forced to admit her backwardness.[19] She was
'one by herself' in the family.

During the first weeks after Catharine's illness, while she
was still 'in arms' and unable to walk, another child joined the
family. On 12 May 1810, William Wordsworth junior was born.
'I am sole nurse. Thank God well and strong.'[20] Just before the
birth, Dorothy received the good news that the expiry-date of
the short Allan Bank lease would not lead to them moving far
away. 'We are to have the Parsonage house', the old Grasmere
rectory, next year.[21] She was exhausted by her attendance
upon Catharine and Mary, and her other emotional strains.
Mary and William were anxious for her to have a well-earned
holiday; the difficulty lay in her obsessive sense of self-sacrifice.
She had taken virtually sole charge of five confinements in
seven years, and regarded herself as indispensable. On 1 June,
she obstinately claimed, 'I could not leave home for more than
one day', and it took all Mary's very considerable tact to persuade her otherwise. The suggestion was that Dorothy should
accompany William to an easier life at the Beaumonts, who
had now moved into their newly built house at Coleorton. 'I
knew she would enjoy herself', Mary afterwards wrote to
William,[22] 'and that she would have been most cruelly disappointed if she had stayed behind & I was also sure that her
heart was set upon going forward, in spite of her wishing to
persuade herself & you to the contrary all this I *knew*.' There
could hardly be a better analysis of Dorothy's complex nature
and her usual stressful battle between self-sacrifice and natural
feeling. The supposedly simple Mary showed rare insight into
her struggles, with the love to resolve them.

Once Dorothy had accepted the plan of a holiday, she grew

calm. 'The country is divine now', she wrote to Catherine Clarkson in June, 'in spite of the embrowned colour of the hills. . . . Our little William is sleeping in my bed and I sit with the sash open looking down to the Lake.'[23] The journey on top of the coach by way of Manchester was so wet that a stream of water followed her along the inn floor, but she was happily no worse the next morning.[24] The month of July she spent with William, in the aura of civilized pleasure which the Beaumonts created around themselves and their guests at New Coleorton Hall. Dorothy saw for the first time the domed hall and galleries completed two years earlier, with the treasures brought from London: the great Rubens, a Nicolas Poussin, three Claudes. The tempo of life was leisurely: breakfast about nine, a free morning, dinner at three, an afternoon's reading, and an evening walk after tea. Later, by the light of many wax candles, they turned over books, prints, and drawings which Lady Beaumont offered for their enjoyment. The lady's maids, traditionally contemptuous of poor guests and small tips, went out of their way to be kind. In four weeks of this charmed existence Dorothy revived, rapidly and strikingly. She was never tired, and looked better than she had done since her first serious illness at Racedown in 1797. 'Her throat and neck are quite filled up; and if it were not for her teeth', wrote William with brotherly frankness, 'she would really look quite young.'[25] Even more remarkably, although insisting there was '*very little hope*',[26] since she might be needed at home to take charge of the children if the maids left, she finally consented to visit the Clarksons at Bury St. Edmunds.

On 10 August, at midnight, she set off on the coach for Cambridge where Thomas Clarkson was to meet her. Coleorton had looked so quiet and beautiful at sunset that she cried bitterly to leave it. Yet once on the road, it became clear that not only her body had been healed by rest. The marvellous gift of living from moment to moment with the intensity of childhood, and offering this experience to any casual reader, rose, after long weariness, in all its old vitality and strength. The letters she wrote to Margaret Beaumont and to William take us with her on every stage of the journey. There was no seat for her inside the coach, but without hesitation she climbed the ladder to the roof, wrapping herself as well as she could in Mary's old blue

coat and a shawl; 'the night was dry and pleasant . . . but I
found it cold enough upon Charnwood Forest. There, by the
light of the moon, and of our Carriage lamps, I now and then
discovered some scattered rocks . . . indeed I am sure, cold as I
was, that I had a much more agreeable journey than I should
have had if I had been shut up there; for the sky was very beautiful all night through, and when the dawn appeared there was
a mild glory and tender chearfulness in the East that was quite
enchanting to me.'[27] They changed horses at Leicester, where
she warmed herself by the kitchen fire of the inn, and at six in
the morning again mounted outside the coach, next to 'a Gentleman's servant with three dogs, two in a hamper and one who
served to keep my legs warm'. By now a kinship bound the party
of travellers. The servants, who of course rode outside, if they
had been 'my Brothers . . . could not have been more kindly'.
At Stamford, she shared her sandwiches with a young woman,
'better dressed than I was', who offered a cake in return—'so I
saved a dinner'.[28] She had always thought of Stamford as a dull
town, yet, in the excitement of the voyage, 'whichever way I
looked . . . I saw something to admire—an old house, a group
of houses, the irregular line of a street or a church or a spire',
all glittering in the clean and sunlit streets. Here they changed
coaches, and at Huntingdon stopped for tea, which would have
been refreshing, but Dorothy denied herself as 'it was too
expensive a luxury'. The evening was cold, and at last she was
forced to get into the coach for the night, recalling the country
sights and scenes outside. 'We had six passengers—one lady
with a bunch of honeysuckles, which added a scented poison to
the hot air, and when I got out of the Coach at Cambridge I
was quite sick and giddy.' They stopped at the great gateway of
St. John's to set down a passenger, the Professor of Arabic.
Morning light 'from a great distance within' streamed along
the level pavement. She thought of William, in his freshman's
days, passing through the same gate, 'and I could have believed
that I saw him there'.[29] In a few minutes they were at the Hoop
Inn, the end of their travels where 'dear Mr C.' awaited her.
This picture of coaching days, so modest and unaffected, can
hold its own for truth with the famous bravura of Nicholas
Nickleby's journey to the North in the company of Mr Squeers.
  Next day, after a good night's sleep, they went early to Trinity

Chapel, where the organ sounded while Dorothy stood in silence before the statue of Newton, 'with his prism and silent face', feeling 'I am sure sublime—though dear Mr. Clarkson did now and then disturb me by pointing out the wrinkles in the silk stockings, the buckles, etc. etc.'. Later they visited colleges and 'walked in the Groves', where Dorothy sought and found William's favourite ash tree. In the evening, she climbed, bravely though a little frightened, to the roof of King's College Chapel, for a view of all the colleges and the town. 'I think', she wrote, 'as I grow older I have more delight in Buildings, and still more for the sake of remembrances of my Brothers',[30] the long-loved undergraduates, Christopher and William.

Her spirits rose even higher as the Clarksons' gig bowled along the road to Bury St. Edmunds. She was overjoyed to see Catherine after a gap of half-a-dozen years. She had missed her friend so much that the white bedroom-window curtains of Eusemere seemed to blow with a ghostly loneliness whenever she passed the house. Now, each day Catherine rode her pony over the heath, Dorothy walking by her side as they talked, Catherine tolerant and liberal over public affairs, witty in private life. Dorothy admired old Mr William Buck, Catherine's father, in whose house they lived, and Georgiana, his other daughter. Bury itself pleased her: the wide, clean square like a village, the overgrown, bird-haunted Abbey ruins, the bells which marked the passing of the hours. William remained somewhat anxious, knowing her temperament. He wrote her a short letter, and warned Mary, 'do not give her to understand, that you have recd. a longer one—this would make her uneasy'.[31] Mary understood only too well what he meant. 'Dear D. how she distresses herself. I wish she would not allow such things to disturb her so much.' The only distress Mary noticed, though, was Dorothy's usual financial worry; 'she tells me that money goes very fast with her . . . and we are *so poor*'. Catherine Clarkson and her sister made frocks for the Wordsworth babies, passed on riding-habits for Mary and Dorothy, and gave, for cold journeys, 'the prettiest Tippet I ever saw'.[32] Dorothy stayed six weeks at Bury, where 'we pass our time very pleasantly . . . chiefly among ourselves'. She did however make one new and lasting friend in the Clarksons' circle, who encouraged her to extend her journey to London.

Henry Crabb Robinson was four years younger than Dorothy. He had known Mrs Clarkson from childhood and admired her matchlessly entertaining conversation, rating her second only to Madame de Staël. His own life was remarkable, for, inheriting a small legacy at twenty-three, he had done what the Wordsworths had found so difficult, and educated himself in Germany. He took a degree at the University of Jena, paying half a guinea a year for tuition and seven shillings for lodgings, and had since been a successful foreign correspondent of *The Times*. In 1810 he was keeping terms at the Middle Temple to enter the Bar. This worldly-wise, urbane character at once appreciated the quality of provincial Miss Wordsworth as a companion, though it was some time before she could believe she was welcome, not only as William's sister, but for her own sake. Then the comradely friendship of this confirmed bachelor became one of her major pleasures in life.

In September 1810 he escorted Dorothy to London. She remembered with gratitude 'our long journey side by side in the pleasant sunshine, our splendid entrance into the great city and our rambles together in the crowded streets'.[33] They walked every morning, went to the British Museum and one evening to Covent Garden. He was, wrote Dorothy, 'kinder to me than I can express. I shall remember him with affection and gratitude as long as I live.' She stayed with Charles and Mary Lamb, whose kindness, too, was 'unbounded'. The two spinster ladies shared a bed, had many pleasant fireside hours together, or went out to look at the great London shops.[34] They were so happy in each other's company that on 19 October Charles wrote to Wordsworth, begging, on behalf of this woman of nearly forty, an extension to her holiday. She was, he said, 'so much pleased with London and . . . so little likely to see it again for a long time, that if you can spare her it will be almost a pity not. . . .'[35]

Meanwhile Dorothy had already made another excursion, long overdue. On 12 October 1810, wearing a new drab-coloured cloak and bonnet for the occasion and of course the Clarkson Tippet, she went down by coach to visit Uncle and Aunt William Cookson at Binfield Rectory near Windsor. She had left them at Forncett, sixteen years earlier, promising to return in perhaps the only conscious deception of her life. Now,

after the brief and hurried reunion in 1802, Dorothy's uncle
'received me most tenderly . . . and seemed to take great pleasure in having me sitting beside him'. Her aunt was admirably
capable and cheerful as ever. The babies she had nursed were
young ladies of angelic 'sweetness and innocence', and their
brother, though 'not so studious as my uncle wishes', was 'a
noble youth' destined for the army. Two junior Cooksons had
been brought specially home from Eton to see their Cousin
Dolly. Most touching of all was the old nursemaid, who had
been years in the family, and welcomed Dorothy with tears in
her eyes.[36] The smiling farmland, the churchyard loud with
rooks, and the hospitable, solid red-brick rectory behind it—all
made her welcome as if time had stood still, as it seemed to
have done for this kind uncle, who had, long ago, introduced
her to his patron, the King, and who himself was to die there at
the same time as his royal master. Hearing the bell tolling
down the valley from Windsor at George III's funeral in 1820,
he said, 'the good old King is gone, may my end be like his'—
and died without a struggle.[37]

The Lambs and Crabb Robinson urged her to stay longer in
London. True, the children at Grasmere had whooping-cough,
but Mary, understanding as ever, had written that the illness
was 'turned' and there was no cause for anxiety. So Dorothy
made plans to stay at least another fortnight. 'I do indeed think
it a pity to go away without having done and seen all. . . . . I shall
like myself the better for it hereafter.'[38] But a letter came from
William; Catharine was dangerously ill. For once, Dorothy did
not pretend, even to herself, that sacrifice was easy; she confessed 'regret and sorrow at parting from my friends, a restless
night and hurry of spirits the next day'. Perhaps William exaggerated. Yet, in a crisis, there was no doubt of her total commitment to the family. She booked the first place in the mail-coach,
left Lombard Street at half-past eight on Monday night, travelling without sleep or rest, and reached Kendal at two o'clock
on Wednesday afternoon. The news was reassuring; she slept a
night and went home by gig next day, 23 October. John met her,
bounding down the field with joy. Although they had coughed
and vomited for weeks, John and small William seemed sound
enough; but the sight of Catharine in her mother's arms cut
Dorothy to the heart. She was almost unrecognizable, 'worn to

a skeleton . . . like a child bred and born in a Gin-Alley'.[39] Nursing her took all Mary's time, so the stout eight-month-old baby Willy fell to Dorothy's lot. 'At first my arms ached dreadfully and breathing failed, but now I can lug him a mile without rest.' Dorothy slept with Catharine, who coughed and vomited all through the night; neither got an hour of undisturbed sleep.[40]

The only treatment for this exhausting and dangerous illness of childhood was 'a change of air'. Thomas and small Dorothy had already been sent to lodge in a cottage high on the fells overlooking Little Langdale and the vale of Brathay. Here the whole family, adults and children with a maid, Sarah the daughter of the house, crowded in for four nights, with November days of bright sunshine and crystal air. The cottage was poor, but at the door 'mountains, hamlets, woods, cottages and rocks'. Here on the first morning William read aloud the Morning Hymn from *Paradise Lost* Book V, 'while a stream of white vapour, which covered the Valley of Brathay, ascended slowly and by degrees melted away'.[41]

That evening William left them and Dorothy walked part of the way with him. Darkness came on as she turned back and she lost her way in the trackless peat moss. She got into a wood, climbed high dry-stone walls, found herself again in the bog, 'and stumbled on, often above the knees in mud, till I came to a cottage'. She had been brave and calm, 'planning what I should do if I were forced to stay out all night', but with her hand on the latch of the cottage door she found she could not speak for sobbing. The wife lent her dry stockings as a matter of course, and the husband guided her home through the dark. They had been searching for her, the children wretched and even courageous Mary 'trembling from head to foot'. 'Well! it is over, and this good is come of it,' wrote Dorothy to Bury St. Edmunds. 'I shall never again go alone in rough places, and on unknown ground late in the evening.'[42] Scarlet fever was in Grasmere and 'Poor William's anxious mind . . . harassed past enduring', so they stayed the next six weeks in a borrowed house at Ambleside.

Dorothy took a serious resolution this winter. Both the Clarksons urged her strongly to let them publish 'The Grasmere narrative', her account of the Green family tragedy in 1808.

Dorothy refused decisively. The events were too recent, publicity might harm the children, and 'I should detest the idea of setting myself up as an Author'.[43] Even by early nineteenth-century standards of decorum, this seems a case of wilful, exaggerated self-denial. At any level of literature women writers were beginning to come into their own. The published verses of Felicia Hemans began in 1808. Fanny Burney had published *Evelina* anonymously as long ago as 1778. Maria Edgeworth began a series of novels with *Castle Rackrent* in 1800. Certainly Dorothy Wordsworth felt herself ignorant compared with her brothers. Yet even Jane Austen, who considered a classical education 'quite indispensable' and herself 'the most unlearned and uninformed female who ever dared to be an authoress', began unobtrusively in 1813 to publish her masterpieces. Why did Dorothy, with comparable gifts, reject the idea so violently? Was it the logical conclusion of self-denial, or simple unawareness of what she wrote?

The children slowly recovered in time for a Christmas dinner party in the kitchen of Allan Bank, while Dorothy sighed for 'a snug house under the shelter of a hill'. By the end of winter she had not had time to open a book, except when the family were in bed, since her hasty return from London. In May 1811, their tenancy at an end, the Wordsworths moved to Grasmere Rectory. The house, built about 1690 by a rector, stood some forty yards west of the church, beside the road and the Raisbeck, where it poured down from Dunmail Raise to empty itself into Grasmere. The village school stood, conveniently for the children, in the opposite corner of the churchyard. Dorothy, as always, was hopeful of the new home; 'if finished according to our desires [it] will be much more like a home of ours than our present house. It will be a large cottage',[44] while their old furniture 'took to its places', altogether, as she wrote in dialect, 'a very canny spot'. They must expect discomfort at first, for the old house was dilapidated; William had undertaken to direct the builders 'and you know how unfit he is for anything of that kind'. They moved in with workmen still splashing whitewash around, while Dorothy sat at one of her heroic labours, stitching two pairs of new trousers for the boys out of the good parts of two old pairs, for John the backs and for Thomas the fronts. Next she worked 'from uprising to going to bed' on Dorothy's

frocks for going back to boarding-school, and baked 'a large
seed-cake to be packed up in her box to treat her school fellows
on her Birthday'. Haymaking began on the glebe land and
with the servants in the fields, Dorothy spent her time cooking
and cleaning, while sharing the 'serious *labour*' of carrying William's books from the Barn where they had been stored.[45] While
Mary took the children to the seaside, she picked gooseberries
and made preserves. By August 1811 she felt 'settled at home'
for the first time and able to 'set to and read'. She and Sara
were both 'deep in a *Novel*' when visitors arrived. The only
trouble of autumn was that De Quincey, perhaps finding Town
End dark, cut down the hedge around the orchard—'every
Holly, Huckleberry, Hazel'. Dorothy, for whom it was a family
shrine, was so angry that she would not speak to him nor meet
him when he called.[46]

Autumn 1811, as at Allan Bank, showed the real drawbacks
of the Rectory. Once again the parlour fire smoked, so they
were forced to sit either in Sara's or Dorothy's bedroom. The
house stood at river level and with winter rains the field grew
sodden and muddy. 'It is no playing place for children,' wrote
Dorothy. Sara, as always, was more forthright. 'If you knew the
bog which it stands in'; when they came in from outdoors 'the
house felt like a well and made us shiver . . . it is without doubt
a hateful house'.[47] Dorothy found consolation. 'On looking
through the window, I see the moonlight mountains covered with
new-fallen snow.' She walked twenty miles without tiring and
danced to the Christmas fiddlers in the stone-flagged kitchen.
It was the finest Christmas Day ever remembered, as she walked
with William on the morning of her fortieth birthday, 'a
cloudless sky and glittering lake'.[48]

There was nothing to show that the New Year, 1812 would
be the most tragic year of their lives. After it all three, William,
Mary, and Dorothy, would never be the same people again.
True, there was one bad day in February, when Coleridge, to
his sons' distress, drove past their door without stopping on his
way to Keswick. Dorothy made no comment, but simply packed
up his German books at his request, and sent them after him by
coal-cart.[49] In April she was left at the Rectory with three of the
children, to spring-clean while their parents were away. William went to visit the Beaumonts in London, where Coleridge

had returned. He intended to demand an explanation. Meanwhile, Mary had taken small Thomas to stay with his namesake, her brother Thomas, on his farm in Radnorshire. At the beginning of May, Dorothy sent a report of her home activities. Although both maids had been ill, she had got the house cleaned, the parlour painted, the furniture 'scowered and oiled', the pictures in William's study re-hung. She was planning 'our 7 week wash', and had ordered coal, flour, and potatoes though 'very dear'. Outdoors, she had a field to plough, hay stored for the cow, butter and a calf sold for a guinea each, and a new pig to buy. 'The place never looked so neat as it does now.'50 Mary, on holiday, found all this somewhat uncomfortable reading. 'Poor dear Dorothy!' she wrote to William. 'I have been sorry that she should have been such a Slave.'51 A further duty arose with the arrival of letters from Annette in Paris. Dorothy sent them on to William and seems to have made a translation for Mary. Caroline and her mother supported themselves by some indoor work demanding constant 'strict . . . attention'.* Mary's tender heart was touched—'constantly in that place for one so young as C is terrible to think of!' She wrote to William with fine generosity of spirit that she thought Annette's conduct 'very dignified & most heartily wish we were rich enough for you to settle something handsome upon dear Caroline . . . God bless her. I should love her dearly & divide my last with her were it needful.'52 William in London attempted to find some way of sending money to Annette through city merchants who maintained business contact with France. Finally Mary's cousin, Thomas Monkhouse, whose 'good affections' Dorothy admired, undertook to 'get letters sent for me into France', advising duplicate copies.

The happiest member of the Wordsworth family this spring was probably Thomas, the 'coaxing Monkey' as his mother called him. His uncle's farm provided endless entertainment for a six-year-old. The men and lads were 'all very fond of him'. He rode all day backwards and forwards with them on the dung cart, or round the fields before his uncle on a horse '& proud was he'. He watered the garden, watched the fishing in

---

* This was possibly a lottery bureau, since Caroline later applied for a licence to conduct one herself.

the pond before the old house, learned to clip a lamb at sheep-shearing, and collected odd lengths of string for inscrutable small-boy purposes. His mother hoped he would not be spoilt, for 'they are all delighted with him & he *is* a sweet fellow'.[53] It was the holiday of a lifetime.

At Grasmere Dorothy was happy in the company of John, Catharine, and Willy, who was just two; on Sunday, 1 June and Monday, she took the small ones into the churchyard opposite, where they ran races and played all the afternoon. On the Wednesday they all came back at nine from a walk and she put them to bed. After half an hour, John called out to his aunt; Catharine had been sick. 'She was lying with her eyes fixed— and I knew what was going to happen.' She sent at once for the apothecary, but he gave her no hope. Catharine's convulsions continued all night, while Dorothy 'prayed for her release in death; for it was plain that had she lived she could not have recovered the use of her limbs, nor probably of her senses'.[54] At last she admitted the truth. 'The disease lay in the Brain.' At quarter-past five in the morning, without regaining consciousness, Catharine died.

Dorothy had to make all the decisions and plan the funeral, feeling it 'a great addition to our affliction that her Father and Mother were not here'. She wrote to William, asking him to travel to Wales and break the news to Mary; but Mary had already guessed the contents of a letter to her brother Tom. William found her pale and worn with shock and deplorable dejection, unable to eat or rest. She felt for handicapped Catharine not only a mother's sorrow, but the defeat of a devoted nurse who has lost a patient. All she wanted was to return home to her other children, though the funeral could not wait for her arrival. On 8 June, wrote Dorothy, 'we all three Sara John and I followed to her grave—She lies at the South West corner of the church yard under a tall and beautiful hawthorn. . . . We have put a small headstone to mark her grave.'[55]

Another duty was to break the news to De Quincey, who was in London. 'I wish you had been here to follow your Darling to her Grave.' If Mary's natural grief was great, De Quincey's was beyond all bounds. He poured it out in torrential letters to Dorothy, surpassing even her own passionate expressions of devotion to William and the children. He wrote of Catharine like a lover: 'On the night when she slept with me in the winter, we lay awake all the middle of the night—and talked oh how tenderly together: when we fell asleep she was lying in my arms—.' When he returned from London to Grasmere, he often spent the night upon her grave, 'in mere intensity of sick frantic yearning after neighbourhood to the darling of my heart'.[56] Moreover, this most innocent of beings left him a dreadful legacy, for grief increased his casual, occasional use of opium to a true and unshakeable addiction, destroying both domestic peace and friendship with the Wordsworths. William, by contrast, celebrated Catharine's going in a magnificent sonnet, 'Surprised by joy, impatient as the wind', one of those mature works which the poignancy of death could still inspire in him.

For her own part, perhaps because of all her practical duties, Dorothy was unusually calm, feeling Catharine's death 'a sorrow in which comfort is found'. She believed Mary would suffer less if she had seen for herself the inevitability of the stricken invalid's death. The child had not lived to suffer lasting harm; her nature had been affectionate and merry, and now 'sweet Catharine is and ever will be the same in our remembrance as when she was alive'.[57] At first she felt she could not leave the surviving children for a single night. 'I *cannot* leave home . . . I *cannot* do it.' The uncertainty of life overshadowed all future plans: 'do let us see each other again', she wrote to Catherine Clarkson in June, 'if it be God's will that we live another year'.[58] Yet in July she went thankfully to see her dear Aunt Rawson, still lively at sixty-seven, and staying with her old friend Jane, in the Marshalls' lakeland house at Watermillock on Ullswater. On the way home she and William ate a picnic with the Marshalls, discovered a waterfall in the woods, and climbed over the top of Fairfield almost in their old style. Her pleasure in small things revived. A pig was given to them 'which was greeted with much joy by the younger part of the Family . . . and it was impossible to say whether the voice of Willy's raptures, or of Pig's fears, was louder'.[59] On 1 November there was a family celebration at Grasmere. Mary's farmer brother Tom married his first cousin, Mary Monkhouse, hailed by Dorothy as '*a perfect Woman*'.[60] Dorothy returned for a second fortnight of shared confidences and schoolgirl reminiscence with Jane Marshall at Watermillock. Life, it seemed, was beginning again.

On 1 December, the evening before she left, she heard from Sara that Thomas was not well and 'that they were looking out for the Measles'. Next morning she took the Penrith coach towards Keswick, intending to stay a few days with the Calverts —but at Threlkeld William met her. Thomas, joyous and confident explorer on life's journey, was dead. Measles, in the cold, damp house, had been followed by pneumonia, an 'Inflammation upon the Lungs', and he had left them the evening before, as he did everything, swiftly. The last words he said were, 'I am getting better.' When Dorothy came to, she would not wait for a chaise. A cart was going towards Grasmere and she lay upon the straw in 'a kind of stupor', William walking alongside. It was dark when they reached home and she was roused by the sight of candles at the door. Mary received her and comforted her 'with the calmness of an Angel . . . I was ashamed of my own weakness; and bitterly reproached myself that I could not bear the sorrow as she did'. The next weeks were a confused nightmare. The three remaining children all caught measles and the household moved to Ambleside for the sake of being near their apothecary. When God 'Spared to us the rest of the little Flock' and they moved home, Mary, who had endured the weeks of anxiety bravely, finally collapsed. She was emaciated and exhausted, 'weeps bitterly—at times and at night and morning sheds floods of tears . . . it must be struggled against or it will destroy her'.[61] Dorothy and Sara did their best, yet 'in spite of all we could do, the very air of that place—the stillness—the occasional sounds, and above all the view of that school . . . that church yard his playground . . . all continue to oppress us with unutterable sadness'. She thought of Thomas 'so particularly fitted to give and receive happiness', and the familiar scene was clouded with pain; 'wherever we look we are reminded of some pretty action of those innocent children . . . my heart fills to the brim when I think of it, and there is no comfort but in the firm belief that what God wills is best for all of us—though we are too blind to see in what way it is best'.[62] The hand of providence, which she had once seen lying over vale and fell to protect its children, was withdrawn. In its place stood a stern and mysterious God, whose commands must be obeyed. 'God's will must be submitted to and I hope', wrote Dorothy to Richard, 'all of us will be enabled to submit as we ought.' Even innocent children were not exempt from judgement. 'The late warnings make us feel daily the uncertainty of their life, and ours.'[63] After this year, she said, 'I never talk of *next year's* plans, but I think of Death.' As she put it to Jane, 'this is a hard trial . . . but God himself knoweth that he hath worse in store for us'.[64] She told herself Thomas was 'happier than we could have made him; and it is but for ourselves that we grieve'. Yet still her heart cried out to him, 'But oh! dear Child, never shall I forget thee', followed by 'that one thought—"I shall never see him more!"'[65]

14. 'The Ambleside Gentry'

William was able to leave the Grasmere Rectory, with its terrible associations and its ominous nearness to the churchyard, because of an unlooked-for stroke of fortune. Thomas had no sooner been buried, than a letter arrived from Lord Lonsdale. It contained an offer of an annual gift of £100, until such time as a post under Government, or a pension, could be secured for the poet. The first draft of £100 arrived on 8 January 1813. It enabled the family to move to 'a most desirable Residence soon to be vacant at Rydale', near Ambleside, and with none of the unbearably tragic associations of Grasmere. About three months later, just before moving to Rydal Mount, William received firm employment under Lonsdale's patronage, the Distributorship of Stamps for Westmorland and the Penrith district of Cumberland. Stamps on legal documents had become dutiable in 1804 to finance the renewed war against France. The job was not the sinecure that has sometimes been represented, and involved hard, responsible work; but for the first time the family knew security   They took the house without even seeing inside, since the outgoing owners, the condescending North family, had their wine still in the cellars, and refused to admit them. Yet they began to hope. On the May Day 1813, when they arrived, in Dorothy's words, 'the morning is bright and I am more cheerful today',[2] for the first time in five months. Besides, 'we are all very thankful for the prospect of an entire release from care about spending money'.[3] Dorothy was also released from twenty years of guilt that they had not been able to pay back the estate of Aunt Richard Wordsworth for William's Cambridge education, which was at last done. Contemporaries, who criticized William for living on Lonsdale charity, did not realize how much this meant to all members of the Wordsworth family, coming at a time of extreme grief and personal crisis.

Catherine Clarkson, sympathetic yet shrewd, had written to Crabb Robinson as a mutual friend in March 1813 about Dorothy's continuing and passionate grief. She felt it misguided, even ungrateful. 'A little child leaves behind it none but pleasurable recollections.' Mrs Clarkson did not share the

Wordsworths' trust in landscape as a consolation. 'I see in the effects of these losses upon them the evil of living entirely out of the world, especially in that Country. I remember the effect it had upon me.' Social isolation, she felt increasingly, made the family self-centred. 'Our friends have no *acquaintances*. They have neighbours, but in their present circumstances they need the sight of *equals* . . . in whose company they must put some restraint upon themselves.'[4] In this, as in many other ways, the move to Rydal Mount was a decisive change. As though to confirm Catherine's worldly-wise judgement, Dorothy's letters that summer of 1813 show already the first signs of returning enjoyment.

Rydal Mount, her home for the rest of her life, was an old farmhouse up a steep lane from the Ambleside road. About 1760, the agent of the Rydal Hall estate had re-fronted it, and the Wordsworths later added an extra storey. As they explored the rooms for the first time they found that they could see the head of Windermere and across the water of Rydal to Silver How mountain. The garden stretched along the hillside in a green terrace walk, where in summer there was always a breeze stirring. They could enjoy 'the Luxury of floating upon still waters on long summer evenings'.[5] In winter, William and John skated every day on Rydal Water. The old walls were thick, the chimneys sturdy, and, as Dorothy wrote to Jane Marshall, the day after moving in, 'the place a paradise'.[6]

In this setting, with the passing of time, their spirits gradually revived. Continual visitors came to see the new house, 'the envy of the whole neighbourhood'. Ambleside, though so near, was a different social world from Grasmere, having been colonized by lake-loving gentry of the 'calling' type. The nearest neighbours were their landlords, the Ladies Diana and Ann le Fleming, mother and daughter, in their grey stone mansion Rydal Hall. They welcomed the Wordsworths to walk in their grounds, and, as the irrepressible Sara said, 'Lady Di pops in and takes a friendly cup with great pleasure.'[7] Inevitably, their social position changed. As poor Mrs Coleridge, who had always considered them outlandish company, wrote in frank bewilderment to their old friend Thomas Poole of Stowey, 'You have no notion how much respectability attaches to them. Their society is much courted.' In September 1813 they held a tea-party for their former Grasmere neighbours, and Dorothy busied herself making cakes and pies in her old style.

She was not entirely easy in their new social life. Endless calling left little time to read. 'I feel' she confessed to Catherine, 'that much of the knowledge which I had formerly gained from Books has slipped from me, and it is grievous to think that hardly one new idea has come in . . . if it were not that my feelings were as much alive as ever there would be a growing tendency for the mind to barrenness.'[8] The pleasures and penalties of their new life crystallized, in autumn 1813, in outings to auction sales to buy furniture for the larger house. Sara and Dorothy had decided this traditional country treat would be 'the very thing' for Mary. The sale was held in a barn 'and we entered into the spirit of it'. William, usually anxious, seemed carried away by the occasion. He 'made a purchase of which he is very proud but it is dear enough—the drawing room curtains with a grand cornice the length of the Room'. He bid hard for two sofas, but was forced to settle for 'a Meat Safe and another Writing Desk' at a different sale. They ended up with 'far more chairs than we know what to do with', for which the devoted household worked seats, still at Rydal Mount, and stayed the sale out to the end. Dorothy's old exhilaration rises in the account: 'the beds were sold by candlelight and we all walked home in the bright moonshine, I with a water decanter and Glass in my hand and William and Mary with a large looking glass . . . very cheap—1£ 13*s*.'.[9]

Yet at home, with 'a *Turkey*!!! carpet—in the dining room, and a Brussels in William's study', memories of the simplicity of Town End called from the past. Were they, she wondered 'setting up ourselves upon the model of our neighbours—the Ambleside gentry'? She imagined Catherine saying, 'are they changed, are they setting up for fine Folks? for making parties —giving Dinners, etc. etc.?'[10] True, William had been to Lowther Castle, where he met a Duchess and '*heaps* of fine folks'. Yet they still worked, still sewed, cleaned, and gardened together in the old way. They had not changed; nothing had changed, she insisted. Was it her friend or herself she was anxious to reassure?.

Life at Rydal Mount was clouded for some years by a haunting anxiety over the health of the surviving children, understandable

after the griefs of 1812, but extreme. 'Our darling little
William', only two years old when his sister and brother
died, aroused panic at every step of his childhood. There was,
Dorothy confided to Catherine, 'no likelihood of another
child', so he remained the baby, especially cherished when 'my
little ones soon will all be big'. Every childish complaint caused
Dorothy 'awful anxiety'. In winter his stomach was disarranged
by mince pies; in summer apples disagreed with him or he
caught 'loathsome ringworm' from the cottage children. When
he had croup in the winter of 1814, William wrote appealing to
Dorothy, who instantly returned from a visit two weeks early to
help Mary with the day and night nursing—'an anxious . . . a
wretched time'.[11] Worse even than her own anxiety were 'my
Brother's unconquerable agitation and fears whenever Willy
ails anything'. A year or two later she even ventured a rare
criticism of her brother for spoiling this youngest child: 'I am
astonished with his babyishness and really his Father fondles
him and talks to him just as if he were but a year old.' At seven,
Willy was 'treated as the little pet'.[13] She maintained she 'had
no disposition to spoil him'. Yet she too was paralysed by fears
of what might happen in her absence if she went away for any
length of time. 'I never—never—so much dreaded to leave home
as now.'[14] The old joy in nature returned in fugitive gleams—it
would never leave her, nor the love of friends—but life, with
'the uncertainty of all things', grew darker in her sight.

If the children were the centre of Dorothy's anxious love, she
also found time for concern over her brothers. Richard, head of
the family, was a reserved, emotionally self-contained man. He
showed concern for his sister, keeping her accounts and paying
out her income. His gift of a pony and side-saddle had
delighted her.[15] Yet he was never effusive in his affection, and
maintained a private life of which she knew nothing. He bought
two small farms near Penrith, but his visits there hardly ever
extended to his family.

Dorothy's letters to this 'curious brother' were increasingly
plaintive. 'We wish much to see you' in spring 1807 became
'impatient to see you' in the autumn. In summer 1809 he spent
some weeks at Penrith, but they did not see or hear from him;
Dorothy wrote in open reproach, 'I cannot express how much
I was grieved.' At New Year 1810, it was 'I shall be an old

woman before I see you again', and 'I am very much hurt'.[16] Four years later, when he was forty-five, one reason for his pre-occupation was explained. 'My Brother Richard', Dorothy told Catherine Clarkson, 'has married his servant—a young woman about two and twenty!!!' A son John was born and William complained to Christopher of 'the disgrace of forming such a connection with a servant, and that, one of his own'. Dorothy, as always, was won over by a child. At Christmas 1815, she and William walked over Kirkstone Pass to visit Richard and his family at their Sockbridge farm. The little boy, at ten months old, she found 'indeed a sweet creature, very pretty and most intelligent and engaging . . . he forces everyone to admire him'. She was glad to like his mother Jane. Though 'nothing of the natural gentlewoman', she was respectable, comely, and 'kind and attentive to her husband'. Richard, happy with his new household, proved 'as usual, very kind and affectionate'.[17]

Unhappily, by March 1816, he was very ill with an abscess on the liver. His Jane nursed him devotedly and Dorothy sent her affectionate messages.[18] Richard rallied, but swollen legs, difficulty in breathing, and fading sight were ominous symp-toms, and his brothers urged him to make his will, especially to secure Dorothy's settlement. Richard appointed his brothers joint guardians of small John. On 18 May 1816 he died; 'we were very thankful' wrote Dorothy to Catherine Clarkson, 'when God had taken him from his sufferings'.

Christopher, the youngest brother, so like Dorothy in looks, promised to have a distinguished career. He had been admitted to Trinity College, Cambridge, at seventeen, won the Latin declamation prize and graduated Tenth Wrangler. At twenty-five he was elected Fellow and sent Dorothy twenty pounds a year from his stipend; 'God bless you, my dear Brother'. She knew he was already 'desperately in love' and engaged to a Quaker, Priscilla Lloyd, sister of Charles Lloyd, of the great banking family of Birmingham; as a Fellow he could not marry, but must wait a few years for a church living. Dorothy remembered Charles Lloyd's quarrel with Coleridge and felt him a 'dangerous acquaintance', though Priscilla seemed 'inter-esting and amiable'.[19] Christopher was ordained at Norwich, where he became private tutor to the Bishop's eight-year-old

son, a domestic post which assured his public career. He became
Chaplain to the Bishop, and, as Dorothy proudly wrote in
1802, 'quite a University Preacher'. In 1804 he was appointed
to a living near Norwich, and could at last marry Priscilla;
Dorothy wished them 'many years of peace and love'. After
John's death, she wrote piteously, 'let us who are left cling
closer to each other';[20] yet their lives seemed to be diverging. A
month later, in March 1805, the Bishop of Norwich became
Archbishop of Canterbury, and invited Christopher to accompany him as Chaplain. Dorothy described them living 'very
nicely in a nice house' at Essex Place, near Lambeth Palace;
Priscilla was reported by William 'a very sweet woman', and
Christopher gave his brother 'books he can spare'. The Archbishop's Chaplain 'kept much to his study' and in February
1810, Dorothy announced 'My brother Christopher's publication', six heavy volumes of *Ecclesiastical Biography*; 'I am
reading it with great interest'. She claimed to find parts of it
'delightful', a considerable test of sisterly feeling. In the same
year, at thirty-six, Christopher 'got another living—600 per
annum'[21] as Dean of Bocking, Essex. The contrast with the
hard-pressed household at Grasmere was marked.

Yet trouble lay in wait for this successful brother. In the first
three years of marriage, he had three sons. 'Tell me if he is at
all like Dorothy or John' was their aunt's instinctive response.
Two more boys died in infancy of 'convulsions'. Priscilla grew
ill and depressed. Though she 'recovered health and spirits'
with her family at Ambleside, Dorothy grew mistrustful of her.
When overworked Christopher failed to visit them at Coleorton, 'I think that surely Priscilla would not let him come
to see us'.[22] When Priscilla again fell ill in 1811, Dorothy
remembered the increasing instability of Charles Lloyd, and
feared the illness 'connected with madness. . . . Christopher
will have little comfort with her and much anxiety.'[23] At
another attack, he sat tenderly and patiently by her bedside for
hours soothing her night after night; 'perhaps Priscilla might
have suffered less if she had had a less indulgent husband,'
Dorothy wrote, family solidarity ousting sympathy. But Priscilla really suffered. In 1815, she was dead. After the still-birth
of a longed-for daughter, she had 'some flying spasms' and was
gone. At the same time her brother Charles Lloyd collapsed in

the 'woefullest agonies of utter depression' and was removed
from home. Dorothy felt 'Death is nothing' in comparison, and
found consolation in Priscilla's end, for 'who knows what this
might at length have become?'—a judgement of dreadful
irony.

Priscilla's parents took charge of the three boys, whom Dorothy grew to love greatly. Christopher was left alone. 'I have
no thoughts of residing with him,' wrote Dorothy frankly to her
old friend Jane. 'I could not give up my present home for any
other.'[24] William agreed for her to visit the widower, but 'his
mind is so bent on performance of his duty and he is so strong
in Christian fortitude, that he does not seem to want anyone to
lean up on, even for a time'. 'My trust is', he wrote simply, 'that
all her sufferings and labour being over she is now among
the blessed.'[25] Increasingly rigid, distrustful of 'Sectaries',
Christopher endured the loss he could never forget, without
complaint or self-pity, continuing his successful career, and his
bleak, solitary life. Dorothy was reduced to asking Mrs Clarkson to tell her if he fell ill, since no one else would. 'People are
always afraid of making me uneasy.'[26]

William's health, too, fed her hungry anxiety. Early in 1811,
she first wrote he was 'troubled with a weakness in his eyes
attended with swelling in the lids'. This has been identified as
trachoma, the 'Egyptian conjunctivitis', an infection spread
through Britain by servicemen returning from overseas wars.
Repeated attacks troubled him with watering eyes, granulation
of the lids, and inability to bear any strong light. Sara, weary of
sitting in the dark, later said he was 'the worst manager of himself in the world', and liable to harm his eyes 'by too much
care';[27] but Dorothy shared his constant apprehension. Like
him, she imagined too easily the 'dark steps' of blindness. To
her usual copying of poems, she added the duties of writing letters at dictation and reading aloud. This was merely an extension of the 'task of my own to perform', self-imposed years
before, to help 'my dear Brother William, and contribute to
make him fit to accomplish the works he meditates. This is an
awful thought and trust me', she had promised Lady Beaumont, 'I will do my utmost.'[28] A new anxiety now was the
effect of Williams's paid employment, with all its responsibilities. Although he had a reliable and trustworthy clerk, Mr

Wilkin, for the accounts, the Distributorship was no sinecure; he was 'too bowed down by excessive cares' to compose for a year and a half. When on 13 April 1817 he wrote, Dorothy reported at once, 'Today he has composed a Sonnet and in our inner minds we sing Oh! be joyful!'[29] Such anxious love for the whole family left her with little peace of mind.

Though the family came first, others too sent for her in times of crisis. She spent the first three months of 1814 at Keswick with Mary Barker, a forthright, indomitable spinster friend of Southey, nicknamed The Major. One might, as Sara said, 'have some good fun' with her, but at present Mary had an anxious task, to nurse a young man of eighteen who had visited her and fallen ill with 'violent spitting of Blood . . . and great weakness'.[30] It was Basil Montagu the younger, the 'little Basil' of happy Racedown days. Dorothy at once gave up all other plans and shared Mary's care for him. The cold was bitter; they could only lift him out to make the bed. Yet by April he was well enough to sit in the drawing-room or ride for exercise, and Dorothy was 'restlessly impatient to be at home again', where William needed her to correct proofs of *The Excursion*. It must be printed, she said, lest it should 'please Heaven to take him'.[31] The publication of *The Excursion*—five hundred copies at two guineas—in August 1814 found her on the defensive. Hazlitt's reviews in the *Examiner* she considered 'more a Criticism upon Country Life . . . than upon the poem . . . he has the audacity to complain that there are no Courtesans to be found in the country'.[32] Jeffrey's famous verdict in the *Edinburgh Review*, 'This will never do', showed he 'played the fool'.[33] She admitted to Catherine Clarkson, 'the effect of the publication has not been such as I expected. I thought that a powerful band of fresh admirers would have been immediately formed.' Yet her own belief in William's genius and his sacred mission was untouched by material success or failure. 'As to the permanent fate of that poem or of my brother's collected works I have not the shadow of a doubt.'[34] A year later, in August 1815, she reaffirmed her faith. 'As to us, *we* shall never grow rich; for I now perceive clearly that till my dear Brother is laid in his grave his writings will not produce any profit. This I now care no more about. . . . His writings will live.'[35]

Her own taste in reading grew more austere. Dorothy's

changed attitude shows in her views on the love-interest in novels. No one could expect her to feel the same at forty-four as she did at fourteen, in the full glow of adolescence, when she read Richardson's *Clarissa* with such enthusiasm. It is a book full of the sensibility current among young people of her generation, the conflict of love, duty, and even sensuality carried along by an enthralling story-line. Yet this was no passing youthful fervour. As late as 30 August 1807, when she was nearly thirty-six, Richardson's novel was still her touchstone for a book of absorbing interest.[36] By February 1815 love in fiction no longer appealed to her. 'When love begins', she wrote to Sara Hutchinson, 'almost all novels grow tiresome.'[37] She had just read the recently published *Waverley*, in which, she adds, 'as usual the love is sickening'. That Dorothy, who had apparently accepted without censure the fervid emotions of Clarissa, should find Scott's respectable manœuvres 'sickening'—that is, tedious—shows a fundamental alteration in emotional matters. Even an innocuous work by Mrs Opie 'made us quite sick before we got to the end of it'.[38]

This austerity of judgement was not confined to the printed page. Inevitably, it applied to life and people also: 'all people are better when they keep away from parties—and their fine clothes'[39] was her verdict on an entertainment at Kendal. Little Dora's self-will, encouraged by 'indulgence or pleasure', must be subdued by 'Rigorous confinement'.[40] De Quincey, the children's favourite, at last did what she had hoped and tutored Johnny every day in Latin; but she was sceptical of his 'nominal hour . . . either the scholar learns with uncommon rapidity . . . or the Master tires'.[41] She had noted in De Quincey the symptoms she had come to condemn in poor Coleridge's sickness; the lassitude, the inertia, the long mornings in bed. 'Notwithstanding his learning and his talents, he can do nothing,' she judged. When he unaccountably lost his gold watch, Sara was prompt with an explanation. 'He doses himself with Opium and drinks like a f.' Worse was to follow when De Quincey fell in love.

He chose a girl whom Dorothy's hand, in passing, had brushed with immortality. On 13 March 1802, a day, she had noted, of 'very hard frost. Little Peggy Simpson was standing at the door catching hail-stones in her hand—she grows very like her mother. When she is sixteen years old I daresay that to her Grandmother's eye she will seem as like to what her mother was, as any rose in her garden is like the rose that grew there years before.'[42] The seasons had passed and now Peggy Simpson was 'sixteen years old'. She had grown up in her father's farmhouse, the Nab, on the shores of Rydal Water, so out of the world that when De Quincey lent her *The Vicar of Wakefield*, she was surprised and disappointed to learn the characters were not 'real people'. It was this total simplicity, like little Catharine's, which gave timorous De Quincey courage to love; but the Wordsworths were irritated by his talk of Peggy's 'angelic sweetness'. Dorothy had perhaps forgotten the little girl in the doorway, for she wrote that Peggy 'to all other judgments appeared to be a dull, heavy girl and was reckoned a Dunce at Grasmere School'. All the Rydal Mount family showed open disapproval. Wordsworth attempted to accompany De Quincey on his night walks to prevent him visiting the Nab, the tall figure and the short locked in comical cross purpose. It was in vain. 'At the up-rouzing of the Bats and the Owls he regularly went thither', Dorothy reported, 'and the consequence was that Peggy Simpson, the eldest Daughter of the house presented him with a son ten weeks ago.'[43] The child was born and baptized at Grasmere in November 1816 and in February of the next year his parents married. In this time of scandal and distress De Quincey hoped for some support, 'an act of friendship so natural and costing so little' from the family to which he had shown years of humble devotion. Peggy was to be a loyal and loving wife through a marriage spent in evading her husband's creditors. Yet the marriage flew in the face of class distinction and chastity, two master rules of the nineteenth century. Neither Mary nor Dorothy, to De Quincey's mortification, called on the bride. Dorothy's verdict was 'Mr. de Quincey is married; and I fear I may add he is ruined'.[44] They seldom saw him again.

Meanwhile, questions of illegitimacy and marriage had arisen closer at hand. For twenty-one years, with one brief interval, war had isolated Annette and Caroline. Now history took a hand in the Wordsworth family affairs. In April 1814, an armistice was signed between the French and the Allies; Napoleon abdicated and was sent to Elba. Dorothy wrote to Mrs Clark-

son[45] to reassure her that although 'encircled by these mountains', she and the inhabitants of Grasmere had not been un-
moved by world affairs: 'every heart has exulted—we have
danced for joy! But how strange! it is like a dream—peace
peace—all in a moment—prisoners let loose—Englishmen and
Frenchmen brothers at once!' Among the released prisoners
was a young French lieutenant, captured three years earlier,
Eustache Baudouin, who knew Annette and through her of the
Wordsworths.[46] While in England, a prisoner at Oswestry, he
had written to Dorothy and as soon as he was free came to visit
them. He returned to France with messages for Annette and
Caroline. The next news, in October 1814, was that Caroline
was being sought in marriage by Baudouin's brother, Jean Baptiste, a Government *employé* in the government pledge office,
the Mont de Piété, and that she applied formally, as French law
required, for her father's permission. Dorothy learnt the news
'with delight for that dear young woman's sake'. Annette had
long ago given up the idea of marriage with William; yet she
kept the name William or Williams, and brought up Caroline
through anarchy and danger 'in perfect purity and innocence',
in spite of Dorothy's fears that French manners must be 'unfav-
ourable to true delicacy'.[47] It was a heroic achievement, hard to
appreciate in Britain. Now, in return, French law and custom
demanded that Caroline should be supported by her father's
family at her own wedding. Annette implored Dorothy to come
and see for herself that Caroline 'resembles her father most
strikingly'.[48] 'I desire exceedingly to see the poor girl,' wrote
Dorothy, planning to bring wedding presents made in
England. Mary encouraged her; anticipation 'makes the thing
ten times as formidable as in reality it is'. Yet she hesitated.
Caroline refused to marry without her, and on her account the
wedding was put off until April 1815, a sacrifice which shows
how sincerely she was wanted. Annette even took lodgings for
her in Paris.

Then, on 1 March 1815, Napoleon landed in France from
Elba, and all was again confusion. Apart from Dorothy's relief
that she herself was not already in France, her chief sympathy
was for Annette.[49] 'I am exceedingly distressed for poor
Annette and Caroline, especially Annette—Caroline is young,
fresh hopes will spring up for her—but her Mother! so near

happiness and again to lose it!' Annette's own letters were not reassuring.[50] Even as she wrote she heard Bonaparte's advance guard entering the city. She concludes, 'Good God, what is to become of us?' Fear for her Royalist relations sharpened Dorothy's long-standing hatred of Napoleon, and her wish that 'he may be crushed at once, hanged and gibbeted'.[51] She had no patience with the liberal theory that the Allies should not 'meddle with the internal government of France'.[52] Even later, when the news of Waterloo reached her, she was disgusted by the English sightseers, who crowded to get a glimpse of Napoleon imprisoned on board the *Bellerephon*. Her indignation boiled over in uncontrolled passion. 'Oh, I am sick of the adulation, the folly, the idle Curiosity, which was gathered together round the ship that held the dastardly spirit that has so long been the scourge of all whom he could injure. *He* kill himself! No he is too much of a coward, and we can be so dull of perception—so insensible to the distinctions between vice and virtue, as to bend—to bow—to take off our hats to him—and call him great—his looks! fancy them filled with magnanimity—but he is not worth talking about—.'[53] Her own passion seemed to surprise her: 'how I got on so far I do not know'; but many shared her feelings.

Admirers of Napoleon such as Hazlitt who, according to Leigh Hunt, wore crape mourning and went about for days unwashed, unshaven, and drunk, were rare. The mild Southey exclaimed, 'What a mean villain is this Bonaparte', and suggested he ought to be handed over to the common hangman.[54] Just over a month after Waterloo, the Southeys and their neighbour Miss Barker organized a bonfire celebration of the victory on the slopes of Skiddaw. Dorothy, with William, Mary, and Johnny (no longer reckoned as one of the children), attended what proved to be a hilarious occasion, partly owing to William's usual ill-luck with picnics. He kicked over the kettle, and as there was not enough hot water to make punch for the assembled company, the gentry only had their rum diluted, while their servants drank it neat. The feast of roast beef and plum pudding developed into a torchlit Saturnalia, but everyone got home safely by midnight after an eight-hour outing.[55]

Now that Napoleon was finally defeated, Annette and Caroline wrote again, imploring Dorothy to come for the postponed wedding, 'all being safe and quiet', though the aunt had announced, 'all our plans are put a stop to'. A strange, somewhat disturbing, period followed. There is no doubt of Dorothy's sincere desire to see mother and daughter; 'this I am determined upon,' she had already written, 'that nothing shall stop me again from seizing the first possible opportunity'.[56] Yet now the opportunity arose, she showed a curious reluctance to go to France.

One difficulty succeeded another. William, Mary, and Sara were on holiday, leaving her in charge at home, the winter of 1815 was bitter, coaches struggled in 'a sea of snow', the children at home might be ill in her absence, the money would be better spent on 'my niece's marriage portion', she needed a gentleman as escort. Yet the Clarksons visited Paris in 1814 and 1818, and neither time did she accompany them. Mrs Clarkson helpfully suggested that 'if Dorothy did not go to Paris', Crabb Robinson was going to France and would bring Caroline to see her family in England. This kindly plan came to nothing, 'having no Friends in London to whom we could with propriety entrust her'.[57] Perhaps Caroline's marked likeness to her father might have embarrassed the Monkhouses or the Lambs. Further imploring invitations came from Paris. It seems possible that Dorothy did not understand the depth of Caroline's longing for her presence, since French and British marriage customs were so essentially different. For her final decision on 15 August 1815 was, 'it is impossible for me to think of going to Paris this year'.[58]

The wedding, twice postponed on Dorothy's account, at last took place on 28 February 1816. William's formal consent was recorded by a Paris lawyer in October 1815. No member of the Wordsworth family was present, though it was many years before Dorothy learnt how much this had hurt Annette's maternal pride. Much misunderstanding and later distress might have been saved if she had felt able to attend, for the legal obligations of a French bride's father went well beyond consent. Caroline had been christened, was married, and would finally be buried under the name Wordsworth. Nevertheless her legal domicile was France, and her marriage therefore governed by the unvarying Articles 144–226 of the Code Napoleon, based on civil law, and quite unfamiliar to most British lawyers. Under civil law a settlement of capital on the

bride was a legal duty if the bride's father possessed 'a sufficient income'. William agreed to allow Caroline thirty pounds
a year, generous enough considering his other family responsibilities, but not a capital dowry in French legal terms. Further,
by another article, 791 of the Code, such financial arrangements could not be renounced. Since William seemed never to
consider attending the wedding of Caroline, who signed herself
A. C. Wordsworth and was described as 'fille majeure de Williams Wordsworth propriétaire', this essential information
remained unknown to him.

In spite of her poverty, Annette celebrated her only child's
marriage with pride. The civil ceremony was held at the Mairie
of the 3 *arrondissement* at noon, and Caroline, still in her white
sarsenet and bridal veil, went on her husband's arm to the con-
secration at old St. Vincent de Paul on the hill[59] in the afternoon, an arrangement which dated from the Concordat of 1802.
A French marriage contract should have been witnessed by four
parties drawn from each family. On this occasion Caroline's
father was unrepresented, but her mother's witnesses included
the widow of a prince, a duchess, a viscount, a marquis, and a
field marshal. They chose the wedding of this dowerless girl to
pay tribute to her mother's heroic career as a resistance fighter.[60]
The Wordsworths did not seem to know of this, but Dorothy
gave Catherine Clarkson a sympathetic account of the fête.
'Thirty persons were present to dinner, ball and supper',
though Annette 'perhaps for half a year to come will feel the
effects of it at every dinner she cooks!' Annette was too proud
to ask the absent father's help and bore all the expenses herself.

Catherine Clarkson still assumed that Dorothy would visit
her newly married niece. 'You ask me if I mean to go to France
this spring or summer,' replied Dorothy. 'I wish it very much,
but William and Mary are unwilling that I should venture so
soon. . . . I do not like to put off year after year. . . . I believe
William would consent provided I could hear of proper companions for the journey'[61]—but no companions were forth-
coming. By December 1816 there was yet another reason to go
to Paris. Mme de Chateaubriand had noted that nobody now
'expected a girl to be a girl on her wedding day', but Annette
maintained proudly that her daughter's 'modesty was her best
ornament'. Ten months after marriage Caroline gave birth to a

little girl, christened after her aunt in England, Louise Dorothée. William stood godfather by proxy, but once again no member of his family attended the celebrations.[62] This was William's first grandchild. Even if there was no open repetition of the rhapsodies which had greeted Johnny's birth, for Dorothy, who loved all the Wordsworth children so much, it must have been a time of deep emotion. Yet she wrote nothing of her innermost feelings. When the baby was ten months old, she was still saying, 'I am determined to go to France. I am under promise to Caroline . . . and besides I have a great desire to see her and all the Family.'[63] Yet instead, she had gone for an affectionate and happy visit of five months to her foster-mother at Halifax. 'I could not resist', she explained to Catherine, the 'entreaties of my excellent (alas! aged) friend Mrs. Rawson.'[64] Even the family at Rydal were surprised by how long she stayed; 'still at Halifax,' commented Sara.

Dorothy, and Caroline's father, might have remained without an account of the Baudouin household, but for a fortunate chance. In May 1817 Robert Southey went off with two friends from 'the Brown Bear or the Black one (his complection is of little consequence)' and hired a packet boat at ten guineas from Dover to Calais, to start a tour on the Continent. By 17 May the party was in Paris. Wordsworth had confided to Southey 'what I already knew', and asked him to visit the Baudouins, without appearing to know the secret of Caroline's birth. Southey wrote to his wife at Keswick a long and confidential account of this interview. He called at the modest apartment in 47 Rue Charlot, an unfashionable quarter, not far from the Place de la République. Caroline was the only person at home; 'she did not know my name and spoke no English'. Southey struggled on in 'my French', and as soon as he mentioned Wordsworth, Caroline replied without hesitation, 'Mr. Wordsworth is my father'. They talked for an hour, Southey feeling he was in a play. 'She wept a good deal and was very much affected during our conversation.' He found her 'interesting', in the contemporary usage, that is, appealing to mind and emotions, 'with much more of natural feeling than French manners and surprisingly like John Wordsworth . . . the little French Dorothy is very like her mother, a sweet infant in perfect health and good humour'.

Next day he returned to take lunch with them and to meet Caroline's husband and mother. Jean Baptiste seemed fond of his wife and little daughter. Annette spoke of him as 'the best of husbands'. Southey shared their meat pâté, Gruyère cheese, and white wine, then Annette—with what distant memories of an Englishman—wanted to make him 'a good cup of tea'. She also 'proposed *punch*' to celebrate the occasion. In spite of their having no English 'we parted sworn friends'. Moreover, the author of *Goldilocks and the Three Bears* had made another conquest; 'I was stopt on the way downstairs to take leave of the baby who had been pleased to smile very graciously on me and take me into favour.'[65] Reading this warm and human account by a mere acquaintance, one can imagine how delightedly great-aunt Dorothy would have been welcomed by the family.

She contented herself with the fact that 'Southey saw Caroline and her Mother, Husband and Daughter Dorothée and was very much pleased with them'.[66] The question remains: why did she wait in all six years before visiting these women so deeply interwoven in her past life and so long loved? Other friends, the Clarksons, Crabb Robinson, were travelling to and from France; Robinson himself noted she was a 'very good traveller'. Did her anxious fears for the surviving members of William's legitimate family, Johnny, Dora, and Willy, somehow outweigh her feeling for his eldest daughter and grandchild? Would Caroline's strong resemblance to the Wordsworth family be a source not of pride but of pain?[67] If love in the pages of a novel now appeared 'tedious' in her sight, how did she see the love which had given life to Caroline? She chose to put nothing in writing, so these questions may never be answered. They leave, though, a shadow of inner conflict and distress.

15. 'Dear Antelope'

If Dorothy Wordsworth's fear and anxieties for her immediate family were sharp, her zest for life was keen in proportion. Scenes, characters, friendship, the world of nature, yielded unfailing pleasure. The events of the day still filled her middle age with the old, youthful excitement. In her late forties, her energies and her enthusiasms were exceptional, and remarked upon by everybody.

The great event of the year 1818 was the Parliamentary election for Westmorland, which took place in June 1818. Electioneering began six months earlier and soon obsessed Dorothy utterly. 'You will think', she confessed to Sara in March, 'that my head is turned by this election, that I can think of nothing else, and true it is.' In politics, William was regarded as the oracle. Half a dozen years before, his wife had written to him 'it is plain there is no one competent to be placed at the head of affairs . . . all here wish *you* were the chosen one & then they are sure things would go right.'[1] The setting was dramatic: a post-war England with twenty-six thousand troops stationed in manufacturing districts, a government using spies, secret police, and secret Committees of Parliament to maintain a slippery grip on public order. Westmorland had always been safe Tory country; the powerful Lowther family considered the two Parliamentary seats almost feudal property and expected their candidates to be returned unopposed. Now this comfortable arrangement was threatened by the arrival of a reforming Whig candidate, Henry Brougham. Brougham had started life as a clever, ambitious barrister and brilliant contributor to the *Edinburgh Review*. Powerful Whig friends placed him in Parliament, where he argued against slavery and for public education and law reform. He was aggressive, eccentric, and determined to make his mark.

William undertook, with Dorothy's passionate support, to act as unofficial election agent for Lord Lowther and Colonel Henry Lowther, sons of his benefactor. Her feelings were soon made plain. Brougham had long been a leader of the Anti-Slavery cause in the House of Commons, and Thomas Clarkson as a former local freeholder wrote to the *Kendal Chronicle* on 'the absolute necessity . . . for continuing Mr. Brougham in the House'. His wife was surprised to receive a sharp rebuke from Dorothy. 'I cannot but lament that Mr. Clarkson has thought it right to lend his help to such a cause—he is little aware of the rebellious spirit stirred up in the country. . . . The majority of the populace of Westmorland are ready for revolution I firmly believe. . . .'[2] Mrs Clarkson later wrote to Crabb Robinson with the freedom of old friendship that though she could not bear to admit it to her husband, finding the Wordsworths 'so thoroughly Torified . . . is a little drawback upon the pleasure of our intercourse even to me'. Crabb Robinson, who admired Wordsworth's poetry, was sorry he 'cannot change with the times' or see 'the dangers of a thoughtless return to all that was in existence twenty-five years ago. . . . I doubt the discretion and wisdom of his latest political writings.' Mary Wordsworth maintained her level-headed judgement. Sara was frankly impatient when a letter came from Dorothy 'all full of election matters. I wish it were over for they are all possessed by it—and, as Mary says, it is pitiable that William should be thus diverted from his natural pursuits.'[3] Meanwhile Dorothy, while dutifully lamenting that 'the misguided mob . . . cry aloud for Brougham',[4] clearly revelled in the excitement of his election meetings. On 12 March, she walked sixteen and a half miles to Kendal in five and three-quarter hours to hear Brougham's speech, which 'would have been sublime if the cause could have been a good one'.[5] Was she in swiftly repressed sympathy with some of the candidate's arguments?

Brougham addressed his election speeches to the public as a whole, including those without votes; he treated local politics as part of national policy, and appealed to many who resented the Lowther dynasty. His tall, bony figure, his harsh yet expressive features, his voice, now thundering and now whispering, gave him the magical power of a great actor. On this occasion he made dramatic use of his own arrival in Kendal during a fierce snowstorm, Dorothy against her will was stirred by the 'Banners, horsemen, music and the immense multitude on foot, all joining in one huzza fearless of the driving storm the spectacle was grand'.[6] Her excitement, however, turned to contempt when the orator referred to William as a 'secret agent', and attempted, unsuccessfully, to quote Shakespeare, an election ploy which Brougham introduced throughout the campaign. Her final verdict on 'the mob'—Brougham's supporters—was 'no joyousness in it but a great deal of odious coarseness'.

Parliament was dissolved on 10 June 1818 and writs issued for a General Election. The result of the polling at the end of June was a foregone conclusion. The Lowther election machine concentrated troops at the polling booths, brought coachloads of sailors from Liverpool at five shillings a day 'to keep the peace', and armed two hundred special constables called '*Bludgeon* Men' by the Whigs. The inns were crammed with soldiers drinking at Lowther expense,[7] and frightening the freeholders. The expenses, in the Lonsdale MSS, appear as: Gross £21,000, £6050 to inn-keeper Appleby, £4250 to other inn-keepers, £3000 for stage coaches . . . £2,800 lawyers, £1000 for music and constables.[8] On 4 July 1818 Lord Lowther and his brother were declared, in the family tradition, 'duly elected for the County of Westmorland'.

Dorothy Wordsworth, once a romantic democrat and friend of passing beggars, regarded this triumph of electoral corruption with the most astounding complacency. 'I really think that our Party have carried themselves with moderation on their triumph,' she wrote to Catherine Clarkson, who may have had very different feelings. Innocently, she related the effect of the election on her family's social position. 'As I say we gained some new Friends': the Norths, formerly rich and condescending, 'give us always a friendly greeting'. Colonel Lowther, painfully shy and 'like a Rustic from one of our mountain vales', called; slowly this unpromising Parliamentarian gained courage to address Mary and Dorothy. Lord Lowther actually came to stay at Rydal Mount for three days 'and my sister and I liked him much'.[9] From first to last she had no doubt of the rightness of the Lowther cause. 'I wish not success to any opposers of the House of Lonsdale; for the side that house takes is the good side!'[10]

Just at this time, a member of the younger generation of Romantic poets had observed, by chance, Rydal Mount in the grip of election fever. John Keats, at the outset of a walking-tour of the Lakes and Scotland, dined at Bowness on 26 June, and was told by the waiter that Wordsworth had been there recently 'canvassing for the Lowthers'. 'Sad, sad, sad,' com-

mented Keats, adding, charitably if inaccurately, 'yet the family
has been his friend always.'[11] Next morning, Keats visited
Rydal Mount, but had another disappointment. William was
away at the polling station at Appleby, where he had taken
Mary and the boys. Keats's walking-tour companion, Charles
Brown, remarked, 'The young poet looked thoughtful at this
exposure of the older.'[12] Dorothy and Dora were out
somewhere, so Keats 'wrote a note for him and stuck it up over
what I knew must be Miss Wordsworth's Portrait',[13] thus
presenting a puzzle since no portrait of her survives from that
time. Keats had already made a pointed reference to William's
inclinations by writing of him as 'Lord Wordsworth'. Though
unwavering in his admiration of the poetry, he went on his
journey disquieted about the older poet's way of life.

The next visitors to Rydal Mount, unlike the roving poet,
were received with elaborate preparations, all made by
Dorothy. They were the Wilberforce family, with six children,
five horses, and seven servants for a seven weeks' stay from
August to October 1818. Arrangements were begun by William
—'but the business was soon given up by William to me and
innumerable were the letters that passed between me and Mr.
Wilberforce'. She rented two houses for them and various beds
in village houses. First came the servants, the housemaid and
kitchen-maid, reasonably one feels, objecting to sleep upon 'a
Mattress' only, and the cook, aggrieved that she could not get
'a drop of Porter' nearer than Ambleside. Southey watched the
topsy-turvy chaos with fascination: 'such a household! The principle of the family seems to be that provided the servants have
faith, good works are not to be expected from them.'[14] The
Wilberforces themselves were, by contrast, 'all joy, animation
and thankfulness' when they arrived. Running to meet Mr Wilberforce, Dorothy was 'much affected'; it was twenty-seven
years since she had seen him at Forncett Rectory and he
appeared worn out by continuous anti-slavery work. It was a
delight to find his mind as lively as ever, and the children happily talking 'all at once'. Mrs Wilberforce was meek and good,
but, as Dorothy admitted to Catherine, called her illustrious
husband 'Wilby', in 'that whininess of manner—tending to
self-righteousness, I do not like'.[15] The young people endeared
themselves by not minding 'a few showers', and enjoying their

9. 'treated myself with a shilling's worth upon the water': Lambeth

10. 'great delight': Brinsop Court

outings.* One little boy even left his umbrella on the coach, to be retrieved on Dorothy's anxious directions—'a Gingham one with a black handle with an Eagle and a Motto under the Eagle carved on the handle'. Details of the spring-fastening took another four lines, for the enlightenment of the Whitehaven Coach Office. With all its hard work, the visit made her deeply happy. 'There never lived on earth, I am sure,' she wrote to Jane, 'a man of sweeter temper than Mr. Wilberforce.'[16]

After this meeting with an old friend came a memorable excursion with a new one. Dorothy and Mary Barker had been drawn together by the months shared in nursing Basil back to health. 'The Major' was a rollicking spinster with an enjoyment of society. She formed a plan, unfulfilled, to visit France in 1816 with Dorothy, 'in high spirits' and 'determined to win a French Marquis'. Sara, when ailing, recovered under her hospitable roof. William even contributed to her poem 'Lines addressed to a Noble Lord', exhorting Lord Byron to leave his evil ways and come to the happy Lake District.[17] She had now built herself a house among the mountains, rocks, and streams of Borrowdale, six miles south of Keswick by a rough road, where she lived cheerfully alone the year round. She had, Dorothy wrote with appreciation and not a trace of envy, 'resources within herself'; she was an amateur painter, enjoyed music, reading, and hill climbing, all in the light of 'a reflecting mind'.[18] In fact Miss Barker, carried away, and 'having', as Mrs Coleridge said, 'a taste for magnificence,'[19] was to find herself in debt to the builders; before a year was out she would have to follow the usual trail of British debtors to Boulogne. This was happily hidden in the future that first week of October 1818, when she and Dorothy Wordsworth planned a great expedition.

This started some miles higher into the mountains than Miss Barker's Borrowdale cottage, at Seathwaite beyond the old lead mines at the head of the dale, thence to go straight up the steep ascent of Esk Hawes, Miss Barker herself having become, like Dorothy, 'an active climber of the hills'. Taking their provisions for the day in a cart to the foot of the mountain, the two

---

* Of the sons on this visit, three became Roman Catholics and one a Bishop of Oxford, a notable opponent of Darwin.

ladies unloaded the picnic to be carried by a local helper, while a reliable shepherd neighbour of Miss Barker was to act as guide. The October morning was brilliantly fine, 'and every little stream tumbling down the hills seemed to add to the cheerfulness of the scene'. They arrived at the top of Esk Hawes exhilarated 'without a weary limb . . . and a sense of thankfulness', Dorothy wrote, 'for the continuance of that vigour of the body, which enabled me to climb the high mountain, as in the days of my youth'.

This, though, was to prove only part of the day's expedition. 'We had attained the object of our journey; but our ambition mounted higher.' On Esk Hawes they were at 2,500 feet, and Scafell itself seemed attainable. They set off, but realized the distance was too great, and they might be benighted if they tried to reach the summit. They therefore settled for the companion peak of Scafell Pike, and were rewarded there with a picnic spot in sensational surroundings, among perfect conditions.

The sun had never once been overshadowed by a cloud during the whole of our progress from the centre of Borrowdale; at the summit of the Pike there was not a breath of air to stir even the papers which we spread out containing our food. There we ate our dinner in summer warmth; and the stillness seemed to be not of this world. We paused, and kept silence to listen, and not a sound of any kind was to be heard. We were far above the reach of the cataracts of Scaw Fell; and not an insect was there to hum in the air.

While looking round after dinner at the tremendous views in every direction, their guide warned them not to linger on the top long, for a storm was coming. They were incredulous, but 'it is there,' he said, pointing out to sea over Whitehaven, 'and, sure enough, we there perceived a light cloud, or mist, unnoticeable but by a shepherd, accustomed to watch all mountain bodings'. They had hardly packed up and set off than 'the tiny vapour swelled into mighty masses of cloud which came boiling over the mountains'. They sheltered under a crag, but the storm passed away almost as suddenly as it came, leaving them hardly wet, and able to enjoy multiple rainbows on the various peaks, while when they regained Esk Hawes, every cloud had vanished.

Coming down by a different route, over a shoulder of Great

Gable, they went by Sprinkling Tarn and Stye Head home down into Borrowdale, travelling in the waiting cart from Seathwaite by moonlight. Their guide had summed up the day as they sat on Scafell Pike, saying, 'I do not know that in my whole life I was ever at any season of the year so high up on the mountains on so calm a day.' The short, dramatic storm had only added to this effect. The sensation of sheer enjoyment permeates Dorothy's description in a long letter written to 'Johnsie', William Johnson, the former curate and master of Grasmere School, now teaching in London. In the manner of the much-earlier journals, written over fifteen years before, she gives her own vital sensations, every sight and sound underscored by some imaginative phrase. On the Pike, 'the huge blocks and stones . . . cover the summit and lie in heaps all round to a great distance, like skeletons or bones of the earth not wanted at the creation'. As a human touch, 'To add to our uncommon performance on that day Miss Barker and I each wrote a letter from the top of the Pike'. One feels that once again, the whole Dorothy is here, free from the constraints of duties, self-imposed or imagined, which, of late years, had so often hemmed her in.

An unexpected visitor arrived in the first week of 1819. The Reverend Dr Andrew Bell, a rich, forceful, and irascible educationist, was already well known to the Wordsworths from his visits to Southey at Keswick in 1811 and 1813. In the 1790s as Superintendent of an impoverished Orphan Asylum in Madras he had devised a system by which selected 'monitors' could teach other children cheaply by rote. The Schools were strictly under Church of England control and for this reason William considered 'the Madras System . . . the most happy event of our time'. He had recommended Johnson, the Grasmere curate, to Dr Bell, who had taken him to London,[20] while Dorothy and Mary had continued to teach 'Madras' each afternoon in the Grasmere churchyard school. Dorothy had undertaken to edit and rewrite 'dear Andrew's' scattered pamphlets as a continuous book, a labour which occupied her spare time for the whole of one summer. Dr Bell had addressed her jocularly as 'dear antelope', adding, 'nobody scolds me as you do' on literary points. He professed admiration for her work, but later threw the manuscript away and re-published his own version. Sara, as always, was forthright. 'He is a queer creature—God knows';[21] but Dorothy, whatever her feelings, gave no sign of author's vanity or resentment.

She welcomed Dr Bell warmly when he arrived, although the house was filled with preparations for 'a grand Ball'. Mrs Coleridge was staying, together with her daughter Sara and Edith Southey, 'two sweet girls'; Dora's school-friends also were invited.

You may be sure we have mirth and merriment enough, with such jinglings of the Pianoforte as would tire any but very patient people. . . . The house turned inside out. Ballroom decorated with ever-greens, a happy employment with hard labour for the Girls.[22]

Dr Bell appointed himself 'the Lasses' Friend' and insisted on staying for the party. He visited the trinket shop to buy presents for the girls and gave each one a guinea; a great wind-fall for any schoolgirl. He also inevitably seized the chance of a ready-made class to instruct Dora's headmistress in the 'System'. The grand ball took place on 7 January, and Dorothy danced with the young people, just as she used to in the kitchen at Town End. Mrs Coleridge, sitting with William, watched her with amazement and lingering unease. 'She looks nearer to 60 than 50 in her profile, owing to her extreme leanness, and loss of teeth.' Sara Coleridge thought 'there was something unnatural in the incongruity of her face and figure together with her extreme agility in the dance'. Mrs Coleridge put it down to Wordsworth's influence; William thought, she said, this 'deep enjoyment in a woman near fifty and his beloved sister . . . was worthy of admiration'.[23] Dorothy, happy as a girl at a party herself, seemed unaware of these undercurrents, and without regrets for vanished youth. 'You will be surprised when you see me in face a perfect old woman,' she wrote frankly to Catherine that summer. 'I have only eight teeth remaining.' A year later she confirmed that she weighed only six stone, twelve pounds, and 'my chin projects as far as my nose'.[24] Yet she did not complain, as any woman might have done. What mattered to her was the fact that 'for my part I have as much enjoyment in walking as when I first came into Westmorland'.[25]

There was more time to enjoy herself now since the dangers of the family's early childhood were safely past. John, at six-teen, 'grows a man almost'; they were thankful Dora (young

Dorothy) at fifteen 'continues a child', though this happy state
of innocence could not last long. 'Our darling little William'
was their remaining link with childhood. Crabb Robinson,
visiting Rydal Mount in 1816, thought the young Wordsworths
appeared 'amiable children if not of superior intellectual endow-
ments'.[26] To their aunt, they were beyond all price. Dorothy
seemed to play an equal part with the parents in choosing their
schools. Dora, 'the apple of Wordsworth's eye', as Crabb
Robinson saw, was the most easily educated. Even as a child
she was 'so clear headed' that she taught a neighbour to play
whist by playing the other three hands herself. At fourteen she
was sent for three years to Miss Dowling's boarding-school at
Ambleside, where her aunt wrote approvingly, 'No hour of her
day is unemployed.' At first tears filled her eyes when she saw
any of her family, but once allowed the 'indulgence' of tea at
home on Saturdays, she grew happier.[27] At Ambleside she
learnt the ladylike accomplishments her hard-working mother
and aunt lacked: sketching, embroidery, 'touching the keys' of
the piano. There was even talk of a Latin master. By the age of
nineteen she was able to help her headmistress, who was very
fond of her, as 'Sole Teacher'. She taught forty children 'pen-
manship', reading, and French, escorted 'the tribe' on walks,
and took care of five or six little ones in her bedroom. Dorothy,
who had always been critical of William's favourite, admitted
handsomely, 'The Children are all very fond of her . . . she
performs her many duties surprizingly for so young a person.'[28]
Unlike her aunt at the same age, Dora would have been equipped
to earn an independent living if necessary.

The two boys presented more problems, but Dorothy was
quite sure what they needed, and prepared to argue even with
Mary on their behalf. John had been for three years at a small
private school kept by a clergyman in Ambleside. She admitted
'his slowness is inconceivable', but like many fond relations
chose to blame the school, rather than admit the problem. She
was determined that Willy, should be better educated. His
father treated him as if he were 'but a year old . . . the little pet
. . . the little darling. I had long been convinced that nothing
but a removal from home could save the Boy from ruin.' Mary
naturally shrank from it, but for once Dorothy would not yield;
'his Mother could not be brought to this conviction till it was

forced upon her'.[29] Their father hoped, vainly as it proved, to get both boys appointed as Scholars on the Foundation at Charterhouse,[30] forty boys, 'the Children of such poor Men as want means to bring them up'. The Foundation offered free board and lodging. They would read 'approved Authors, *Greek and Latin*', have Latin prayer morning and evening, and '*Greek* Testaments for their use in the Chapel . . . as are read in the best-esteemed Schools'. There was also the prospect of nearly thirty closed Exhibitions at Oxford and Cambridge and Church preferment to follow.[31] John was entered first in October 1819. His mother and aunt sewed hard at the new suit and new shirts required by the rules, and enquiries about 'country woollen stockings in winter' were addressed to a perhaps surprised headmaster. Dorothy, improbably hoped the ancient foundation might be 'upon the Bell system', and prayed for Johnny's purity at this most anxious parting: 'God grant that he may preserve his ingenuous dispositions!'[32] Alas for their plans; after so much agitation, John was ineligible since he had passed his fifteenth birthday. Dorothy reacted proudly to the disappointment. 'My Brother would not for the world send him to any other *public* school,' she told Catherine firmly, 'he intends him for Cambridge', and the boy would study 'under his Father's sole tuition'.[33] As Crabb Robinson remarked, not unkindly, 'she has an enviable habit of forgetting all that is unpleasant in what she loves'. In the end John was sent to Sedbergh, where, to the school's credit and his, he at last made good progress.

Meanwhile Willy, at nine and a half, had already left home. In spite of his father's extreme anxiety about his health, such was family belief in Dr Bell that the little country boy was sent in October 1819 300 miles to the Central School of the 'System' in the crowded, dingy streets east of Grays Inn Road. Here he learnt the commandments and the catechism by rote, among a limitless number of boys. At least he knew his headmaster, Mr Johnson, with whom he lodged, and the Lambs entertained him at home 'to a Bullock's heart'. Charles Lamb wrote Dorothy, who was 'very doleful after parting with poor Willy', a kind and tactful letter about the boy, who 'tho' no great student hath yet a lively eye for things that lie before him'.[34] In 1820 she would go to London and take him to Charterhouse. Relating these public-school plans to Catherine, she broke off. 'It is time to dress for dinner, so I must stop.'[35] Can this be the family which sat, discussing life and literature so eagerly, round the high-tea table in the kitchen at Town End? It evidently can, for when times change, people change with them.

The original plan, to live near a good grammar-school, would have been happier for the children and much cheaper, but this was by common consent forgotten. Even with Wordsworth's extra income, three children at different boarding-schools represented a considerable expense. It is not clear if Dorothy contributed to the fees, but she at once related them to the nagging problem of her eldest niece Caroline; 'We have no money to spare for me to spend in France.'[36] Thus the disturbing question of William's French daughter was shelved for another year. Meanwhile, in the spring of 1820, she had other, unavoidable expenses. Her remaining teeth were increasingly painful, and after much family discussion she took the decision to have them removed and false teeth fitted. Once decided, she faced this considerable ordeal with courage and cheerfulness.

On 29 April 1820 she went to London, where Thomas Monkhouse, Mary's cousin, took infinite pains to help her find a good dentist; 'it does me good to think I have such a kind friend near me,' she wrote appreciatively. After a fortnight's enquiries, they settled on Mr Dumergue, 'the best in England', although his fee of fifty guineas worried her at first. Afterwards, she felt it had been well spent. He extracted eight remaining teeth and stumps, of course without anaesthesia. 'The tooth-drawing was not half so bad as I expected,' reported his patient, 'though bad enough.' Determined to see the best in every friend, she added what must have been an unusual compliment. 'He is certainly a delightful operator.'[37] She expected as soon as the gums were healed to be fitted with a complete new set; meanwhile she could not show herself even at church. The artificial teeth of the period consisted of two rigid half-hoops, with carved bone, tusk, or sometimes sheeps' teeth, held in place by a hinged steel spring which forced them hard against the wearer's gums. The owners of these doubtful luxuries were frequently seen to take them out at table, in order to eat without pain.[38] Almost equally distressing to the sensitive was the traditional jocularity they aroused. Charles Lamb, the kindest of men, dwelt on 'Eating, Talking, Biting & c. but not gnashing—gnashing of Sea Horse

teeth must be horrible', until Dorothy was forced to turn away
with 'averted looks'.[39] It is a measure of her habitual self-efface-
ment that no word of these discomforts entered her letters
home, which were filled instead with family news.

She knew that Mary's chief anxiety would be for news of
Willy, who had been away from home for six months. Although
she had not reached London after her long journey until eleven
at night, he was brought to visit her the very next morning, and
they walked out together, Willy 'completely overcome with
pleasure. . . . You can judge of the joy of our meeting.' She set
to at once on her usual duty of mending his clothes, and went
with him 'at his first going to the C. house'. The old priory
off Clerkenwell Road had become a collegiate building, with
hall, chapel, and lodgings round a small, crowded quadrangle,
shared between the foundation boys and eighty 'decayed gentlemen' in the cloak of the charity. The four hundred paying
boys, such as Willy, used the open space before the gate, now
Charterhouse Square, as a playing-ground.[40] To the fond
aunt, it seemed a strange world. 'The boys were at play without hats and he and I wandered about the large Square like two
forlorn things—nobody noticing us.' Did she feel an instinctive
anxiety for cosseted Willy in the jungle of an old-style public
school, the daily diet of classics as unvaried as boiled meat,
bread, and bullying? It seems perhaps she did, for in her usual
way she instantly denied it; 'no better Boy can be—and he will
take care of himself amongst the 400'.[41] She would not see him
too often, 'for fear of unsettling him'. At their final parting on
15 May 1820 the ten-year-old boy broke down and 'sobbed
aloud; but though my very heart was melted I checked his
grief'.[42] They parted 'chearfully', leaving Dorothy to dream of
John at Sedbergh and distinguished academic careers for both
nephews.

Her happiness was completed by the fact that she was staying with her brother Christopher in his rectory at Lambeth.
Lambeth, though larger, was still essentially the separate township on the south bank of the Thames, surrounded by fields
and orchards, where Gilbert White had heard the green wood-
pecker laugh in the woods near Vauxhall. Dorothy's window
looked over the trees of the Archbishop's park, where the
thrush and the blackbird sang in the long shadows of evening.

Everything here depended upon the Archbishop of Canterbury's palace, and Christopher, as his Chaplain, was fully occupied; '*his* occupations are constant and except once he has never been able to go out with *me*'. His three boys were at school, but Dorothy was not lonely. Crabb Robinson, calling, found her with Lady Beaumont, 'a fine old lady and very lively', but bitter against Brougham and the Whigs. She dined out and met Miss Joanna Baillie, the Scottish poet, 'one of the nicest of women', whom she praised interestingly as 'without the least mixture of the literary Lady'. She walked to the Monkhouses, navigating her way with a map from Lambeth to Queen Anne Street. She visited the Lambs, and Crabb Robinson devoted time to her with pleasure, in 'a delightful walk through the park' or 'a delicious stroll over Westminster Bridge'. The nearness of the river was a perpetual delight. She walked in the palace garden, 'by a wall down which you look upon the water, which is for ever varying with boats perpetually enlivened . . . in complete seclusion, yet looking out on the busy scene. There I have walked before and since sunset.' With childlike pleasure she even 'treated myself with a shilling's worth upon the water', while reproaching her own extravagance.[43]

Christopher, though 'as kind a Brother as can be', remained preoccupied until, in mid-May, he fell ill, with headache, fever, restless nights, and rapid pulse. Dorothy said, truthfully, that it grieved her very much, but she instantly dropped all other activities and seized the chance of nursing him, writing his letters, cancelling his engagements, watching him day and night, and recording his physician's opinion. It was two weeks before he said he was 'getting on; but very slowly',[44] and she stayed through June to guard his convalescence, happy that he needed her; 'neither I think', wrote Sara Hutchinson, 'will D. be happy to leave her brother C. unless his health is perfectly re-established'.[45] It was 23 June before she wrote to Dora of leaving, and sending presents, not forgetting the three Rydal Mount maids, Mary, Betty, and Mary Anne, who each received black silk à la mode for bonnets. William and Mary had arrived in London, and the three together shared her last enchanted view of the river at night: 'we stood under the trees of the Palace gardens by the water side . . . and looked upon the brilliant moonlight water scattered over with Boats and adorned with

hundreds of golden pillars—the reflection from the lamps'.[46]
She had no fear of London at night, and no moralistic condem-
nation of the city compared with the country. Free for once
from duties and worries, she yielded to her unique gift for joy
in the passing moment.

The visit to London and Lambeth was only the prelude to a
larger expedition. All the children were now safely and, she
hoped, healthily at school. There was no obstacle this summer
to going, safely escorted by her own family, on the half-desired,
half-dreaded expedition to the Continent. A reason was found
in the marriage of her cousin Tom Monkhouse to a Miss Horrocks, and their honeymoon tour of Europe, accompanied by a
Horrocks sister and a maid. The Wordsworth family party consisted of Dorothy, Mary, and William, quieted in his anxieties
over the boys, and with his stamp duties handled by deputies.
On the morning of 10 June, they took coach to Dover, where
they joined the newly married couple and proceeded to Calais.
There they bought two stout carriages to transport all seven to
Belgium, Germany, Switzerland, Italy, and back via Paris and
Boulogne. At Berne, they left the carriages, hired another,
and, near Interlaken, a charabanc, which would take the whole
party of seven, and, as Mary noticed, cast 'a curious shadow'.

They crossed the Wengern Alps by mule, though Monkhouse's bride, finding the tour too strenuous, returned with her
sister to wait for the others at Berne. Crossing the main body of
the Alps on foot, by the St. Gothard, they wandered about the
Italian Lakes, walking, driving, on horse, or by mule, and in
the third week of September reached Geneva, where they
found Mrs Monkhouse, Miss Horrocks, the maid, and their
carriages awaiting them. They were in Paris on the homeward
journey by 2 October, and ultimately were in London by 9
November, just four months after setting out.

There is no lack of information about their journey. All the
travellers apart from William, who had visited many of the
places thirty years before, kept diaries, Mary and Dorothy extensive ones. So, in the second half of the tour, did Crabb
Robinson, who met them by previous arrangement at Lucerne
in August. If the picturesque Trelawny is to be trusted,
Dorothy was observed at Lausanne looking, with Mary, like
'pedestrian tourists, fresh from the snow-covered mountains,

the blazing sun and frosty air having acted on their unseasoned skins, as boiling water does in the lobster'. Used to the Lake District, Dorothy revelled in the freedom of the even-greater mountains. When she first saw the Alps near Zurich, she gave, as Mary said, 'a scream that made us think something had happened', having only just realized that the huge shapes were mountains and not clouds.[47]

It is a pity that so little of such delighted shocks of surprise gets into the huge journals she kept for later copying. As she wrote it all up at Rydal Mount the following winter, she felt herself the tedium of 'copying that enormous journal, which I can never expect anyone except a few idle folk ever to read through'. She herself thought it 'all written about the outside of things hastily viewed'.[48] For once her habitual self-criticism seems justified, and it is even confirmed by the usually admiring opinion of Crabb Robinson. Preferring her Scottish journal, he designated her Continental one 'a hurried composition not much better than my journals', which he did not value.[49]

It is perhaps an irony that the moment when Dorothy seems most inspired by the excitement of the tour in foreign and unfamiliar places is at its outset, when the party left Calais, and 'we jolted away as merry as children—showed our passports—passed the gateways, drawbridges, and shabby soldiers, and, fresh to the feeling of being in a foreign land, drove brightly forward, watchful and gay'. Her feelings were probably associated, more than she knew, with William's. She was, after all, setting out from the town of Calais which, eighteen years before, in 1802, had been the scene of so much vital shared emotion for them both. Then, too, looking back a dozen years more, the whole Continental tour was, in her mind at least, a retracing of William's footsteps in 1790, when he had walked with Jones through Revolutionary France and Switzerland, and she, left behind at Forncett, made it her task to 'trace his paths upon the maps'. Now, in her own tour journal thirty years later, she reverted to the emotions she felt then and 'my brother's wanderings thirty years ago'. In fact, the picture of what she herself used to be persistently colours the present journal, and, where it does, sharpens intensity. On the road to Cologne and the Rhine, 'I felt as much of the glad eagerness of hope as when I first visited the Wye, and all the world was fresh and new'. Reverting to her earlier, energetic self, as the carriage 'rolled along (ah! far too swiftly! and often did I wish I were a youthful traveller on foot)', she was unconsciously drawn back to 'the shapeless wishes of my youth'.

Probably the moment when her journal best shows the shock and immediacy of earlier impressions is her description of the party's first experience of an avalanche on the Jungfrau.

We were all on foot, and (at the moment when, about to turn to our left and coast along the side of the hill, which, sloping down to the base of the snowy mountain, forms a hollow between) suddenly we heard a tremendous noise—loud like thunder; and all stood still. It was the most awful sound which had ever struck upon our ears. For some minutes, we did not utter a single word: and when the sound was dying away exclaimed, 'it is an avalanche!'.

Only when they had heard the sound several times more was she able to particularize. 'The sound is loud as thunder, but more metallic and musical. It also may be likened to the rattling of innumerable chariots passing over rocky places'.

Otherwise her impressions often bear out her own letter to Sara Hutchinson from Interlaken, 'everything that is most beautiful in this country reminds us of our own'. As with the Scottish journals, comparisons with the Lake District abound, and are continually repeated. A Rhineland bridge is compared more than once to 'a one-arched Borrowdale bridge'. The 'greenish hue' of the waters of Lake Thun 'is much less pleasing than the cerulean or purple of the lakes of Cumberland or Westmorland' and lacks their picturesque groups of cattle; high commendation is given to the Vale of Chamouny which 'transported us instantly to our vallies of Westmorland'. Typically, 'Let no one speak of fatigue in crossing the Alps who has climbed Helvellyn,' she wrote to Catherine Clarkson.

Throughout these weeks of travel, though weary from the ceaseless rattle of the coach, the constant packing and unpacking, and for a fortnight at Coblenz ill 'from the state of my bowels',[50] Dorothy faithfully kept up her journal. Nothing was too fleeting or too small for her writer's eye: the hills above Lake Lugano 'wrapped in green soft glowing light without shadows' or the Bernese women's caps, with black gauze wings, 'like butterflies'. Beside her, Mary wrote a similar, though shorter, record. When the party reached Paris, both these journals abruptly take a new turn. Mary, after brief remarks on the heavy rain, the 'dirty streets', and the neglect of Sabbath observance, which 'made me melancholy', concludes: 'I shall here close my imperfect notices. Paris Oct[r.] 2nd.'[51] Dorothy's entry was still more decisive: 'here ends my Journal. From this time I took but occasional and brief notes . . . it would have been useless to arrange them and attempt describing things, lately so well described in books that are in everybody's hands.'[52] Clearly, they had agreed to conceal the existence of William's French family in journals which friends would read. Was this to protect his poetic mission from cynical gossip? Or did Dorothy's puritanical standards, which had condemned De Quincey so harshly, now shrink from illegitimacy in her own family? Not a word remains in writing to say openly what she felt.

Memoranda Paris in the back flyleaf of Mary's journal record lodgings in the Rue Charlot, the Baudouins' street, and the weeks from 2 to 28 October, when they left by diligence, spent in ceaseless sightseeing. Eustache Baudouin, the only English speaker, escorted his brother's literary father-in-law round Paris, the ladies keeping up as best they could. Each day had its expeditions: the Luxembourg, the Louvre, the Jardin des Plantes—where Wordsworth preferred the animals to the Louvre statues—Versailles, Notre Dame, the Catacombs, the Invalides, and the manufactory of what Mary reasonably thought were called Goblins.[53] Dorothy's chance notes mention only an hour at the 'Mont Pièté . . . very curious', presumably to see Jean's place of work, and the cemetery she called 'Père Chèse . . . as Madame says "trop embelli" for a receptacle of the dead'.[54] 'Madame' remains unexplained. Could it have been a personal cipher for Annette? For the fact remains, that despite all the exhausting sightseeing, October 1820 in Paris was essentially the long-delayed visit for which Caroline had pleaded since 1814. They were there to see William's daughter, last tenderly remembered as a little girl delighted by dancing lights on the seas at Calais, now the mother of a family. An unknown husband must be met, and—most poignant of all— Annette, the survivor of so many dangers, must be encountered once more, and met by Mary.

On the first day 'we are all to meet in the Louvre at one o'clock'. Thus Mary and Annette met for the first time in the most public place imaginable, midday in a famous picture-gallery. They could not speak a word of each other's language, but both had dignity and generosity to carry them through the ordeal. What Dorothy felt at this fusing of past and present in her own life, the revolutionary ideals of youth with the strict morality of middle age, remains by her own choice unknown. Next day, 3 October, the whole party went to 47 Rue Charlot, the Baudouins' apartment. Crabb Robinson was impressed by Annette, who seemed clever, and by her record of heroic resistance to tyranny. He liked everything about Caroline, 'except that she called Wordsworth "father" which I thought indelicate'[55] In fact Annette, as she later revealed, cherished in Caroline precisely the character she had inherited from her father, 'la ressemblance morale qu'elle a avec lui'. She was also proud of William's grandchildren, Dorothée, the engaging baby who had charmed Southey, now a little girl of nearly four, and a second baby Anne, who bore the name of her paternal grandmother. A third daughter, born in 1823, would try to study her famous grandfather's works; one volume of the 1815 edition of his *Poems*, of which they were unable to understand a word, was still in the Baudouin family in the 1920s.[56] This was still there as late as 1937, when Malcolm Muggeridge and Hugh Kingsmill visited Madame Blanchet, Wordsworth's great-great-granddaughter, in a quiet flat in Paris; there was also a pencil sketch of the poet, which he had given to Annette. The French family still treasured the connection and kept evidence of it with pride. Though puzzled why William had never married Annette, Madame Blanchet felt there was no bitterness on her part, nor on Caroline's, who was remembered, incidentally, long after her death for her gaiety and vivacity. The loyalty of William's French family and their frank pride is attractive. Dorothy though was never to see any of them again.[57]

16. 'We all want Miss W.'

Once on the homeward journey, Dorothy's spirits appeared to rise. Both she and Mary briefly took up their journals again. At Boulogne, though within sight of England, they were detained for ten days by storms and contrary winds, but on 3 November 1820 enjoyed a 'Delightful walk . . . saw *Dover Castle* and the beloved cliffs of England'. They used the delay to visit Mary Barker, who, like most of the English community in Boulogne, was trying to pay off her debts; meanwhile she supported herself in cheerful independence as a carver and gilder of picture frames 'to admiration'.[1] She looked not an hour older and Dorothy joyfully felt in this friend's house that 'we were already seated by an English fireside'. Even the maid, 'Borrowdale Agnes', kept her good Cumberland speech. Unknown to Dorothy, though, Mary Barker surveyed her friend's devoted household critically; at her next meeting with Robinson she described the poet as 'spoiled by having three wives—that is two beside his real wife . . . his sister and Miss Hutchinson', a sharp remark.[2]

After near-shipwreck on the Boulogne shore in rough weather, at last they sailed and on 7 November 'with thankful hearts we set foot upon our dear native land!' On the coach journey to London, Dorothy kept exclaiming with joy over fields, cattle, pretty cottages. They rattled over Westminster Bridge—'an interesting moment'. Did it recall her last returning journey over the Thames with William in 1802, haunted by melancholy and tender thoughts of the child Caroline? Dorothy wrote to the Baudouins; letters and a packet containing British needles and razors, then the best in the world, were sent, via Crabb Robinson, as presents to Annette. Letters to her from Dorothy are known, dating from 1821, 1830, and 1831,[3] though others, now lost, may well have been sent. As from the earliest days in 1793, the duty of letter-writing was left to Dorothy.

After a few days in London in November 1820, the travellers took coach from Fetter Lane to Cambridge, whence on 2 December, Dorothy set off in the ten o'clock night coach for Ipswich, to spend Christmas with the Clarksons. She had not confirmed her date of arrival, but the Clarksons, knowing her habits, were only mildly surprised when she appeared unannounced at nine in the morning. She had walked the four miles from Ipswich, through gently sloping farmland and the small park, overarched with trees, which sheltered Playford Hall.* The Clarksons had moved here five years earlier and Dorothy had not yet seen their small domain. Playford was a fortified manorhouse, built late in the fifteenth century, its buttressed walls surrounded by a moat, and its courtyard entered by a former drawbridge. Ivy crept over walls and roof, and in summer the arched windows were curtained in honeysuckle.

Dorothy had persistently and anxiously considered her old friend an invalid, and was amazed to see Catherine now, up at eight every morning, brisk and busy. Playford was quiet, a village of two hundred and fifty souls, its only notables the perpetual curate, the constable, one farmer, and the village shopkeeper. Even the nearest inn was five miles away.[4] Though so quiet, it was never dull. The two women visited sick and needy cottagers, and walked or drove out in Catherine's little gig. When it rained they were 'happy by the fireside . . . in my dear little Parlour', where Dorothy had her favourite sofa seat. Above all they could talk freely. Dorothy had promised 'more . . . when we meet' of the carefully anonymous 'Friends' in Paris; to Catherine, who had known of them since before William's marriage, she could speak freely of 'the History of the Family' and her own exertions. Catherine, sympathetic about this, was critical of the rigid standards of Rydal Mount and the Wordsworths' reverence for the landed gentry. 'I am vexed', she confessed to Crabb Robinson, 'when I hear of vulgar Tory sentiments coming from those lips which utter so much wisdom.'[5] In Catherine's company Dorothy for once heard liberal arguments and was honest enough to find contradiction stimulating. If money allowed, 'how pleasant it would be at least once in every two years to have the rouzing up of a journey Southward!'

Happiness was complete on 14 December when Willy arrived on the Safety Coach, in charge of the guard, for his Christmas holidays. His aunt was delighted with his vigorous looks, old

---

* The Hall was the seat of the Felton family, one of whom had assassinated Charles I's favourite, the Duke of Buckingham, in 1629.

Mr Clarkson smiled benevolently on him, Mrs Clarkson and her lady's maid spoiled him delightfully. Willy gave no trouble. His favourite game, ever since Dorothy had made him a little scarlet spencer out of Captain Luff's old militia tunic, had always been soldiers; he would happily command imaginary regiments out of doors for two hours at a time, even when snow threatened. Christmas and Dorothy's birthday were affectionately celebrated, and by 9 January 1821 she had returned to Cambridge taking Willy to spend the remainder of his holiday with his uncle and cousins. Christopher Wordsworth's career had taken a forward step in summer 1820, when his lifelong patron, the Archbishop of Canterbury, recommended him to the Fellows as Master of Trinity College, Cambridge. After his illness in London he did not expect to live long and his signature in the College Register was, he said, 'Like the hand of a man signing his will'. In fact he was only forty-six, forceful and active. As Vice-Chancellor he embarked on a vigorous campaign of reforms: extending the College buildings to save undergraduates from the moral dangers of lodgings, and attempting to broaden the curriculum. Dorothy arriving at the Master's Lodge, with its Tudor drawing-room, great staircase, and historic portraits on the panelling, found her brother's duties occupied him 'every day and almost all the day through'. After the intimacy of nursing him at Lambeth, she 'never saw him from breakfast until dinner time'. In the Lodge, 'all is so quiet and stately within and without doors' that 'I am much alone but solitude is not irksome'. The winter weather was clear and she 'walked the Groves through many a time' or turned over 'noble prints' of the Fitzwilliam Museum housed in the Perse School. Christopher was not a man to exchange confidences, yet he admitted he was 'lost in this huge house by himself'.* His three orphan sons came home from school to a holiday tutor each morning and a father who retreated to his study every evening from tea-time until prayers. Finding an aunt who would walk with them, talk with them, even play with them, the boys 'talked so feelingly of their solitude' that she could not bring herself to leave them until the end of the month.[6] Her consolation was family pride in her distinguished brother's

---

* The Lodge, containing the present Fellows' Combination Room and Parlour, was larger then than it has since become.

success. Unhappily this success was not unclouded. Christopher
was a rigid disciplinarian and, like William, an unbending
Tory, while many of the College Fellows were Whigs. In Hall,
the Master was seen choking with passion when a Fellow proposed a toast to a leading Whig politician who had been a
Trinity man. Worse followed when the Master attempted to
enforce compulsory chapel attendance for tutors and undergraduates, angrily demanding the resignation of an assistant
tutor who questioned the wisdom of this rule.[7] Thereafter the
Master lived secluded in the Lodge, seldom entertaining or
receiving Fellows except as business demanded. Even his official portrait was described by his growing sons as 'amazingly
glum'. He was a lonely man, who neither invited nor permitted
pity.

Inevitably it seemed, once reunited with her family, Dorothy
began to worry again. She asked Tom Monkhouse 'how the
Poems and the *Excursion* sell', and fretted that Willy was too
much excited by his cousins and 'the notice of undergraduates'.
He brought and broke a '*gold pin*' for his shirt at 3*s.* 6*d.*, which
a kindly young man replaced with 'one for ten shillings!'
Worse, he told a schoolboy fib about his coach fare to get an
extra half-crown, thus incurring 'the great sin of lying'. By
comparison with his three studious cousins, he 'never once
looked at a Book for pleasure . . . and for his own good I often
wished that the holidays were shorter'.[8] Willy did not; on the
morning of returning to school he was pale and miserable.
Trained on Dr Bell's principle that 'the poor should not be
educated overmuch',[9] it is hard to imagine how he survived the
weekly public recitation of Greek and Latin verses which was a
time-honoured Carthusian ordeal. Early in the term, he was
worsted in a fight and got a black eye. Some realistic doubt
must have crossed his aunt's mind, for on the way home to
Rydal she visited Johnny, and admitted a 'lingering wish' that
Willy were not at such a 'frightful distance in case of sudden illness', but at Sedbergh with his brother to protect him.[10]

Early in February 1821, after ten months away, she was
home at Rydal Mount. Family scenes, family ties, family concerns closed round the returned traveller. At nearly fifty, she
was by the standards of the time, an elderly woman. The family
circle, which had held such profound significance and beauty

in the eyes of Coleridge and De Quincey, was now a group of ageing people, living a fond but restrictingly narrow life. Their troubles: Mary's worry over her absent boys, Sara's loose and painful teeth, William's inflamed eyes and the accident at Lowther when, fumbling with a portmanteau and an umbrella, he was thrown from his horse, all caused Dorothy 'wretchedness and anxiety . . . extreme'.[11] The literary world was shut out from her interests. Coleridge's October 1821 selections in *Blackwood's* from his correspondence with men of letters she dismissed as 'obscure enough to unlearned readers like me', but grew angry when William's *Ecclesiastical Sketches* were reviewed by 'a person who could give such a *senseless* criticism'. She complained the publisher 'has used my Brother very ill'.[12] The distinguished *London Magazine*, edited by Keats's publisher, John Taylor, she described as, apart from Lamb's Essays, 'that abominable magazine'. Perhaps most significant of all, since it concerned an old friend, was the affair of the exiled black Queen of Haiti, who, after her husband's defeat and death, took refuge with the Clarksons at Playford. William and Sara wrote a jocular parody of Jonson, 'Queen and Negress chaste and fair', which Dorothy sent to Catherine, apparently quite unaware that it might give offence. It was only after five months' silence that she had doubts and wrote showing unease that the 'foolish rhymes' on 'poor fallen royalty had . . . been unpleasing to you'.[13] Even general reading lapsed, since William's eyes could not tolerate print for more than a quarter of an hour and in winter he could not bear the light of a candle.[14]

In this family circle, with the children grown up, Dorothy had to find her own work. She chose, with persistent, almost obsessional self-sacrifice, the role of nurse, housekeeper, secretary, and slave to any household in sickness or trouble. 'We all want Miss W.', as Sara Hutchinson put it.[15] The pattern of Dorothy's life in the 1820s was long periods of selfless toil, interrupted by the determination of old friends to give her rest and enjoyment. Both satisfied her deepest necessities, the longing of the inner orphaned child to be needed, and to have her own place within the family. The first family to ask her help at this time was in extreme distress. Edward Quillinan, a lieutenant on half-pay in the 3rd Dragoon guards, was an admirer of Wordsworth. In spite of a 'short and stiff' reception from the poet, he settled in lodgings at Loughrigg with his wife and small girl. Dorothy found them pleasant neighbours. Trim and charming, though little taller than De Quincey, Lt. Quillinan was a clever man and a scholar, willing to take the Wordsworth family for drives with his gig and horses. After the birth of a second daughter Mrs Quillinan had a breakdown; but she made a good recovery and in spring 1822 the family moved into Ivy Cottage, below Rydal Mount. Here, after only five weeks, her clothes caught fire and she was grievously burnt, over a large part of the body. Her husband was called to London on urgent business, and Dorothy volunteered to stay with the sick woman. She dressed the wounds, sat up with her at night when she was restless and feverish, measured out her quinine, and sent encouraging reports to the husband. 'The sores give her less pain—and are going on perfectly as they ought to do.' Inevitably during the second week infection spread, a terrifying sight for a home nurse. 'When the wounds ought to have begun to heal—they grew worse'; the fever mounted and at the end of a fortnight the patient died. 'It was my lot to attend her death bed,' wrote Dorothy, grateful that William and Mary were on a visit to Lowther, and Sara at Harrogate, so they were 'spared the last awful scene'. Edward Quillinan did not arrive until six days after his wife's death, and Dorothy was left to make all the arrangements in the melancholy house. She took the two little girls into the household at Rydal Mount until an uncle took charge of them, chose a plot in the graveyard, 'in the same corner where lie our little Catharine and Thomas', and followed the body to the grave.[16] Afterwards she undertook the business of subletting Ivy Cottage, made a calculation of the accounts, and advised the widower on his outstanding bills,[17] though she admitted finding 'money the most tiresome subject to me in the world'. All this was willingly done for a comparatively recent acquaintance. Her only reward was 'satisfaction to me to find how well I had been able to go through the trial'.

By chance, the Clarksons came north for a long-promised visit two days after Mrs Quillinan's death and accompanied Dorothy to the funeral. Catherine thought 'Miss W. was *uneasy*' during a sermon which dwelt on the dead woman's—fallacious—claims to royal descent, but there was little room for liberal opinions at

Rydal Mount. The Wordsworth family united to condemn 'a sad want of zeal in the district', and when Lady Ann le Fleming built a new chapel of ease at Rydal, the poet hoped for congregations 'highly gratifying to the munificent Foundress',[18] in which his household dutifully joined. Dorothy evidently shared the family respect for rank; her response to the news of their old friend Walter Scott's bankruptcy was scandalized— 'How *could* it happen that he should have so entered into *trade* as to be involved in this way—he a Baronet!'[19] Fears for Scott's income and cherished possessions were apparently secondary to this sense of social outrage.

One of Dorothy's happier sparks of independence showed at the proposal of a mountain tour in the Tyrol for Crabb Robinson and her brother. 'I ventured to utter a thought which had risen before and been suppressed in the moment of rising— "how *I* should like to go".' When William spoke of the expense, however, she was at once 'completely checked. . . . the journey would be impossible.'[20] Remembering her first shock of delight at the Jungfrau in 1820, she confessed in a later postscript, 'what would I not have given to have heard the Avalanche with you!' Then putting the pleasures of travel resolutely from her mind, she once again took up her duties. The distress of Mrs Qullinan's death and William's recent accident were hardly over when Dorothy was drawn into her familiar role of nurse to the Wordsworth children. Dora had already left school and they were all 'happy to have her at home'. She was not a beauty, nor an intellectual like young Sara Coleridge, but she was 'thoughtful, steady and womanly' with 'nothing left of her boisterousness . . . in manner or deportment'. Like her aunt before her, the spirited girl had been subdued. Inevitably, there was one anxiety. Dora had left the crowded boarding-school bedrooms with persistent colds and feverish sore throats. Her aunt complained of her 'dullness and heaviness—that inactive appearance', but was convinced that regular exercise would cure it.[21]

This was nothing to the alarm which greeted Willy's return for the holidays in July 1822. Even his relatives had been forced to realize that he would never make a Foundation Scholar. He was too backward for the work, too spoilt for the discipline, and he frankly hated school. He came home 'with death on his

countenance—a dry fever', which William, more alarmed than
Dorothy had ever seen him, pronounced dropsy. Mary and
Dorothy nursed Willy day and night, hourly expecting his
death. Yet he improved and, when the apothecary forbade
serious books, became noticeably more cheerful, sitting up in
bed to play whist with a jolly old lady visitor of seventy-three.
In a month he was out of danger, but all plans for his education
were abandoned; 'his strength was not equal to the demand
made on it . . . he cannot be sent to school again'. Willy was
twelve when, apart from some rowdy classes at Ambleside with
poor Hartley Coleridge as reluctant usher, his schooling
ended.[22] He stayed at home, the object of his aunt's tender
fears—'God grant there may be no organic disorder!' Officially
he was 'preparing for Oxford' with his father; in reality he
enjoyed the country pastimes of 'drowning five kittens and
catching a few score of fish'. The situation was rich in irony.
Willy, cheerfully unencumbered by intellectual luggage, lived
to be a hardy seventy-three. Dora's sore throats marked the
onset of tuberculosis which ended her life years before the
death of either parent or her aunt.

Dorothy's one holiday of this exacting year 1822 seems, on
the face of it, of a somewhat unrestful nature. It was a return to
Scotland, this time with only the companionship of Joanna
Hutchinson, Mary's youngest sister. Undertaken at a later
time of year than seemed wise—the two women did not set out
until 14 September—it had other obvious disadvantages.
Although Joanna was nearly ten years younger than Dorothy,
she was the invalid of the Hutchinson family, a martyr to
rheumatism, to which the rigours of travel in a country almost
as rough as it had been twenty years before exposed her. She
was also, as she had always been, subject to nervous fears;
Dorothy had nursed her through 'nervous' illnesses in 1803
and 1808.[23] These were no qualifications for a travelling companion. She was laid up by rheumatism after only a fortnight,
and Dorothy had to conclude her own so-called holiday by
nursing Joanna for five weeks in Edinburgh. Also, Joanna's
alarms at the rough company they encountered were excessive,
although in some degree justified. On one occasion, at a toll-
house farm where they were obliged to spend the night, six
miles out of Moffat, they met men who 'in a loud voice with a

very broad Scots accent demanded a glass of whisky' and 'had a rattling wild air and demeanour which would have completely upset us had they overtaken us on the road'. The men indeed left; but Dorothy's sleep that night was disturbed by alarms when she heard the landlord's brother moving out of doors to open the nearby toll-gate for the road traffic.

I was struck with the horrible thought that he had got out of bed to admit the two men who had come in after us in the evening. . . . I thought what if they then settled upon murdering us!

It was, she wrote, 'five minutes during which I was more terrified than ever before in my whole life', and Joanna was equally scared.[24] In spite of the beauties of scenery, it can hardly, with such episodes, have been a comforting holiday.

The next year, 1823, found Dorothy on duty at home as housekeeper to the family, while William and Mary visited the Beaumonts and made a tour of Belgium and Holland. Unluckily, during the five months they were away, Dorothy caught 'a most severe cold' and developed influenza. She attempted to carry on, but by May was forced to admit, 'I am strangely weakened.' When her strength 'fell away' Dora proved 'a tender nurse and faithful housekeeper',[25] and the illness had one happy result since it confirmed the growing affection between aunt and niece.

Then followed an exhausting summer, with large parties of visitors to tea, to dinner, and to stay until the house overflowed; fifteen visitors with their relatives and friends were housed in cottages rented by Dorothy.[26] She insisted she had recovered her strength and wrote in October to a friend in Edinburgh that she 'could now attack Arthur's Seat and the Braid Hills as lustily as when you and I trudged together' on the Scottish visit. It was November before she could at last get away. A gig was sent to take her over Kirkstone Pass to Penrith, where Mrs Rawson was staying during the Races. It was a joy to see her foster-mother, '78 and as chearful and gay as if only 16', walking out in spite of her lameness and attended by a kindly husband. 'The pair seem not to have been touched at all by the lapse of seven years' since she had last seen them. At Penrith she also visited brother Richard's widow, 'a good creature and I have a great affection for her', and had various tea-parties with her own old schoolfellows and their families of 'pretty young ladies'.[27] The brief holiday ended with a visit to Lowther and a long day's walking with William, making their way over the fell from Hawes Water to Rydal; 'I never spent a more remember-able day, seldom a pleasanter', and she was 'neither stiff nor tired'.[28]

November 1823 was marked by another family event, the birth of another grandniece, then called simply 'niece', to Dorothy. Marie Baudouin, youngest of Caroline's daughters, grew up to take pride and interest in the works of her distinguished grandfather, but her great-aunt unhappily never saw her. Even her name does not appear in Dorothy's correspondence, though whether by deliberate suppression or loss is impossible to say. Nor does the early death, in 1825, of the middle grandniece, little Anne Baudouin, whom Dorothy had seen, as a baby, in Paris. There is no written evidence to show how deeply these events—so far in distance yet so near in blood—affected her. Only one fact emerges from the letters which survive. She wrote to Crabb Robinson, 'knowing my delight in travelling', about longed-for trips to Switzerland, the Tyrol, even that sacred ground Rome. Yet she never proposed another visit to her relatives in Paris,[29] either with the family or alone.

From February to May 1824 Dorothy had one of her rare 'rouzing up' journeys to the South, where a series of invitations awaited her. After a month with the Beaumonts she, Dora, and William stayed with Thomas Monkhouse and his wife at 67 Gloucester Place, Marylebone. Monkhouse was ailing and William in 'fits of dolefulness' with a cold and 'sparing his eyes by keeping them shut upon the sofa'. In this difficulty the invaluable Crabb Robinson walked the ladies about to see the sights: the Diorama, the Swiss models, the Mexican curiosities—with a 'live Mexican'—and Dorothy with cloak and umbrella 'paddled' to the new church in York Street for a Good Friday service.[30] A connoisseur of sermons, she went twice to hear the fashionable preacher Edward Irving. In print he had struck her as '*worse* than a Methodist Rant', but in the pulpit his dramatic manner and emotional voice overcame her—'one essential I give him full credit for—*sincerity*'.

After the excitements of London, Cambridge was almost lonely, with 'no companion in this large and quiet house' but

Christopher's son John, now an undergraduate at his father's college and deep in books. The river Cam, though, offered its own distractions: after days of heavy rain in May, 'the Flood' covered the Backs up to Trinity Library. She found it 'very amusing', that is, interesting, to climb Castle Hill when the sun shone out again, and see the waters resembling 'one of the lake-like Reaches of the Rhine'. She could not, of course, be so near Ipswich without visiting Catherine. By 28 May 1824 she was happily settled at Playford Hall, but she could only stay briefly. By the end of the month she had given up the prospect of the visit to the Hutchinsons in Wales and was on the coach to Rydal. Inevitably, she had found another friend in need of her care.[31]

Captain Luff of the Loyal Mountaineers Militia and his wife Letitia had been friends of the Wordsworths from their earliest Dove Cottage days. William and Dorothy stayed at their house in Patterdale during the 1805 journey round Ullswater. Dorothy was surprised to find 'glass jugs all in style, Mahogany tables and dumb waiter' in this mountain paradise, which its mistress dismissed as 'a little pottering spot'. 'Mrs. L. will be doing,' she noted, and sure enough, at Mrs Luff's insistence, the Luffs emigrated in 1812, to take the Captain's gout to Mauritius. Once arrived, they gave 'a wretched account of the place': living ruinous, landscape barren and 'overrun with insects continually biting'. Dorothy sent them William's *Excursion* as a consolation, but before it arrived, the Captain was dead. She imagined what she herself would have felt in Letitia Luff's place, the brooding upon her loss, 'the Pangs of Remorse' for having persuaded Captain Luff to take the fatal journey. 'I wish I could see how she could be made happy,' she worried. She applied to Lord Lowther to secure Letitia a widow's pension, but even his powers could not arrange it. 'With her pleasure in giving, her pleasure in spending she will find herself poor,' lamented Dorothy.[32] Meanwhile Letitia, returning to England in some style with the Governor's suite, showed little sign of brooding. 'She has set her heart on furnishing a cottage,' reported Dorothy in amazement that a solitary woman should make such an independent decision for herself. She at once gave up her own holiday plans and met the widow in June 1824 so that they could travel northward together. 'We shall travel with our

Family cares,' she wrote to Crabb Robinson; 'the whole of
Mrs. Luff's living stock, three singing birds of gay plumage
brought from Mauritius.'[33] Moreover Letitia announced herself 'a bad traveller'. Once arrived at Rydal she was 'very
poorly' and Dorothy could not leave her until she was nursed
back to health.

Once recovered, Mrs Luff found a delightful house under
Loughrigg, Fox Ghyll. She determined to have it at once, with
'all the risk and trouble of ousting the De Quinceys' who happened to be living there. Dorothy anxiously advised economical
lodgings to no effect. Letitia had a flair for houses and gardens,
unlike most women the control of her own money, and all
through 1825 was in full creative flight: 'daily visiting her
grounds' or 'setting gardeners at work', drawing plans and
'overlooking workmen'. 'She will soon be at the bottom of her
purse,' thrifty Dorothy prophesied. By 1826 Letitia was
installed. 'What a pity she should have spent so much money
there! She *would* do it,' wrote Dorothy, profoundly shocked.
Letitia survived the Captain for thirty years, happily 'making
improvements' and provoking her friend's final verdict: 'It is a
good thing for women in general to have a *master*, and Mrs.
Luff should always have had one . . . for she *will* have her own
way.'

While Mrs Luff's affairs took their spirited way, 1825 at
Rydal Mount was saddened by the death of 'our good and dear
friend' Thomas Monkhouse. He went to Clifton, since his long
illness grew no better, and his end in February was 'a merciful
Dispensation of Providence'. Dorothy had never taken for
granted the kindness of this busy London merchant, of whom
she wrote, 'He is in no danger of being so wedded to gain as to
shut out good affections', and his going left a gap in the family
which time could not fill. Writing to his brother John Monkhouse, she recalled 'tried and faithful Friends of the Old
Times', as far back as the old maiden lady—his wife's aunt—
who 'was my good Mother's bridesmaid and I believe dressed
for me the first doll I ever possessed'.[34]

Spring brought both fine weather and spring-cleaning: 'Dust
& labour among the books . . . washing carpet', 'carpet shak-
ing etc.'. In a house with three country servants and a handy-
man, she still insisted on doing this heavy work. The summer

was crammed with guests, both calling and staying. 'We never in our lives had so many visitors,' wrote Dorothy, rather wearily. A list in the back cover of a small notebook includes eighteen distinct families. The weather in August was a 'damp oppressive heat' which tired her, and her gardener's spirit was tried by 'a plague of caterpillars in gardens'. Walter Scott's son-in-law Lockhart, visiting for the first time, thought that she looked ill—'yellow as a duck's foot',—though another first-time visitor, an aspiring young writer from Manchester, Maria Jewsbury, noticed Dorothy's vitality in spite of her age, 'green vigour with grey maturity'. Among the stresses of housekeeping came another appeal for help in an unexpected voice from the past. On 16 July 1825 Thomas De Quincey—so long dismissed by the Wordsworth family as 'the Opium Eater . . . poor little man'[35]—wrote Dorothy a pathetic letter. He and his wife Margaret had now four children; evicted from Fox Ghyll in favour of Mrs Luff, they had been forced to crowd back into Dove Cottage. Struggling with 'the wretched business of Hack-Author', complicated by mountains of debts, he was forced to leave home to earn money, though never enough for their needs. Margaret, lonely and depressed without him, had threatened suicide and 'if that should happen . . . I shall never have any peace of mind or a happy hour again'. He begged Dorothy to go over and drink tea at Dove Cottage, giving his poor wife the relief of talking over her distress with a friend. He reminded her of the bitter year when Thomas and Catharine died. 'Oh! Miss Wordsworth—I sympathized with you—how deeply and fervently—in your trials 13 years ago—now, when I am prostrate for a moment—and the hand of a friend would enable me to rise before I am crushed, do not refuse me this service. . . .'[36] Apparently it was one and a quarter years before Dorothy could bring herself to make the visit for which De Quincey pleaded. She was at home until February 1826, but there is no surviving letter to De Quincey and no mention of his family in her journals. She was then away until November 1826 before she called. This, so unlike her usual impulsive rush to help friends, suggests doubt. Did she understand from his letter how much the Wordsworths' scorn of his marriage had wounded De Quincey? Or was it painful to see between this despised couple the signs of tender and lasting love?

The past continued to haunt her, including an unusual event, a letter from 'Coleridge . . . dear Coleridge'. Acting as sole housekeeper in November 1825, while William and Mary were with the Beaumonts, she invited Hartley Coleridge to dine. All the family had protested at his drunken and disorderly habits when he came to Ambleside to teach, but the school had failed, he was now struggling to 'write for Magazines', and memories of his childhood arose in her. Gratefully he enjoyed a rare good meal, 'joining our young people in a merry game at cards', and walking with them all 'to the Waterfall during a sunny gleam'.[37] During this autumn Dorothy had little time to read more than the *Travels* of Charles Waterton in South America and the *Memoirs* of Madame de Genlis with their disturbing picture of French manners and morals. At Rydal 'we get no new publications'; moreover, she admitted, 'I have not written a word since August.' Looking back, 1825 seemed 'the most bustling summer ever remembered'.

On 10 February 1826 at a quarter to eight she left at last for her postponed holiday, 'always ready', as Sara said, 'to enjoy whatever was proposed—never making any difficulties'.[38] The proposal was a three months' visit, accompanied by Joanna, to Thomas and Mary Hutchinson on their new farm. They had moved a year earlier into the calm, dreamlike country between Hereford and Hay-on-Wye. At their rambling, moated, many-chimneyed farmhouse, Brinsop Court, which could comfortably shelter fourteen 'Cumberland and Westmorland souls', Dorothy passed months of the most unclouded happiness in her life. The evidence for this is her almost wholly unpublished journals, in eleven worn and shabby notebooks, beginning late in 1824.[39] Even in the bustle of Rydal Mount she had found time to note, with the directness of earlier years: 'Larch Green Waterfall', 'Pearly lights on lake', or a sunset: 'purple & ruby hues in sky & lake—Emerald Isle—Gipsies on hill'. Now Brinsop offered fresh delights, with leisure to write of them. She arrived on 17 February to find flowers already out in the hedges and Tom's orchards in bud, although on the Black Mountains snow still lay: 'arrived about five—Tea upstairs all happy and comfortable'. The fireside was enlivened by four children, fond of play, yet brought up with a firm common sense that amazed her: 'warm or cold it is all one to them—they run about and play lovingly together—never creep into the fire —not a word of complaint do you hear . . . perfectly healthy'. She decided Mary was 'the best manager of children I ever saw',[40] and the bond between the two women was strengthened when Dorothy successfully nursed her through a threatened miscarriage. Some of the old griefs still lingered. 'To Church in cart . . . old Mr. Jones beside his children's Graves . . . Girl of 16—would never be well till she was gone to her Maker.' A burdened old woman, a blind child seen at Hay-on-Wye, found their way into the diary. Yet the views revived her spirits. On 2 March, '32 cattle feeding as one on slope opposite my window. Birds singing but sound drowned by grating of mill. Swans floating before me.' The water was 'clear as a lake. Drops make silvery rings.' Behind the house lay Credenhill Camp, a sharp conical hill with prehistoric fortifications, which the boys showed her. It became a favourite walk with the children or alone at sunset.

On 1 April Dorothy planted sweet williams round the farmhouse, and the season for excursions began. She went in the gig to Bredwardine, to the church where, half a century later, Francis Kilvert would be vicar and the churchyard where he would lie: 'Beautiful spot on this bright day . . . celandine— violets—primroses—budding honeysuckle—full blown palms.' She walked round the gardens of manor-houses: Foxley with its cedar and cypresses, Italian garden-house, and 'Demoiselle' dragonflies, or Armiston Hall with its wonderful pear tree— 'Pear blossom in glory'. On 24 April she and Joanna left Brinsop after breakfast for a walking tour down the Wye Valley. They went by Goodrich Castle—'Ruin . . . decaying pleasure grounds—lovely spire in north ruin—moat dry around Saxon tower'—and on to Whitchurch where Dorothy climbed the rocks for a 'beautiful opening of the Wye'. Next day they went on to Monmouth, where they put up at the Angel Inn near the church, crossed the Wye bridge, and took 'beautiful walks'. They were only eight miles from Tintern Abbey with its memories of lost youth, but Dorothy, who could walk forty miles in a day, wrote firmly, 'Tintern given up on account of distance' and beyond this no word. They started the return journey at six next morning and were home for tea on 27 April. The three months of the planned visit were already up and Dorothy was usually impatient to get home, but this time, as Sara wrote, 'She is very much charmed with her quarters . . . & is in no hurry to leave them'.[41]

Instead, she went on 2 May to stay with Mary Hutchinson's brother, John Monkhouse, who farmed at The Stowe, near Hay. On the way she saw a 'Lovely prospect of the Wye—a fisherman in his coracle—one wading'. In holiday mood, free from duties, she spent 'All the morning reading Sir William Temple', and on an excursion with Joanna to Worcester 'saw cathedral . . . talked of Coleridge', though she does not record what they said of him. Her solitary walks were 'as lonely as among the Forests of Germany—Peeps of the snow and blue mountains' in Wales, a shepherd and his dog herding sheep, 'as good as a play . . . a never-ending concert of Birds & now & then the chearful Cuckoo'. She visited the ruins of Clifford Castle, home of Fair Rosamund, a place of evocative beauty, with the 'River secretly murmuring under the old Towers'. In June, she was back at Brinsop Court in time to ride with Tom Hutchinson to see the Hereford election. By July and August, her journal had settled into the purposeful rhythms of farm life.

July 5   Lovely bright moon at 5 . . . carrying hay resumed after dinner.
      7   Returned through fields—wheat almost ripe . . .
     11   After Dinner all play till bedtime—a lovely sunset . . .
     14   Cold evening—Fire in sitting-room. Tea & to bed . . .
     19   Dine beside pool.
     20   Got wet in Meadows by Wye.
     24   Sate in harvest field . . . very hot clear day . . .
     29   Walked in cool meadows with Mrs. H to meet children.
     30   Hot walk to church . . . Flowers fading in Lady G's garden . . .
August 1   Still carrying wheat . . . waited long for gleaners. Very dark in returning . . .
       5   . . . after dinner alone to Foxley—through shady hollow with masses of sunlight—squirrels rabbits hares pheasants . . . on through park-like grounds & woods to open grassy cliff-like Terrace—home at 8 o'clock—glorious prospects.
       7   Lay on the grass.
       8   Hereford Races . . . home by old oak—pretty sports there . . . Great delight . . . Blackberries—nuts—mushrooms—strewn apples & c.

10   Slight autumnal air—gleam sunshine . . . In the even-
      ing with T & M & Girls to Oak Trees . . . though quite
      hollow to appearance solid. Within a family might
      lodge . . .
12   Beggar Boys gathering apples . . .
19   undressed for the night—moon staring me in the face . . .
28   Hop picking begins . . . Mrs. H. and I went to field.[42]

In September Dorothy went for a farewell visit to John Monkhouse at The Stowe. They went to a sale at Weobley Parsonage
—'melancholy appearance of strangers gathered together &
furniture spread on lawn'—while the bells rang in the new
vicar. They talked of farming and were 'busy with Books'
—perhaps accounts, since Monkhouse joined in. The evening
of 8 September, her last at Brinsop, was beautiful. 'Lovely
bright moonshine—walked in garden—moon reflected in pool
—the one white Swan.' Next day she left after seven months.
She rose at six and climbed into the cart for Hereford, very
nervous of the frisky horse and unable to speak for the pain of
parting. 'I neither looked nor spoke. . . From top of hill looked
back on old Friends for last time.'

From Hereford she was going to Coleorton, by way of Worcester, Leamington, Warwick, and Stratford-on-Avon. At
Worcester she stayed with a cousin of Lady Beaumont, entering into her life of good works: 'church services—visits among
poor houses—governess perfectly happy in her situation'. The
visit to the county gaol had its macabre fascinations: 'to the
prison . . . Tread Mill—solitary cells—dark cell—one woman
a fortnight in dark but wd not ask pardon—another man prayed
to be flogged next day . . .'.[43] The coach towards Ashby de la
Zouch was noisy—'Glasses of Ale bring out the native Character'—but the final walk to Coleorton showed 30 September as
'the loveliest of days'.

October with the Beaumonts passed swiftly and happily. She
read the *Examiner* out of doors in 'quietness and brightness',
drove in the woods, walked after dinner with Sir George, lost the
dog, picked 'an autumnal bouquet', talked of William's poems
and Christopher's letters on Charles I as the true author of the
Eikon Basilike. Every evening, wax candle-light and ranged bookshelves offered the rarest of all treats—unlimited reading. In one
month she read Spenser's *Faerie Queene*, Coleridge, Cowper,

Crabbe, Charles Lamb's *Essays, Tristram Shandy, As You Like It, Hamlet*, volumes of 'old plays', 'Irish Letters', perhaps Spenser's, William's poems of course, and on Sundays—when secular literature was forbidden—studied 'Townsend on the *New Covenant*'. On the last morning of her stay the sun 'rose like a golden Ball—flashing light to west'. She dined, said goodbye to 'our dear & excellent Friends',[44] took coach to Liverpool, fighting off clamorous rapacious porters, changed to the Kendal coach, and at last reached home on 4 November 1826 after nine months away. 'I cannot express', she wrote, 'how happy I am to find myself at home again.'

Change and rest, as always, brought calm to her mind. Soon after reaching home, she at last made her long-delayed call on Margaret De Quincey and on 16 November 1826 wrote a report to the anxious husband. She described the children, much grown since she had last seen them: William and Horace lively schoolboys, the girl 'stout and well' in the kitchen, and Mrs De Quincey 'seated by the fire above stairs with her Baby on her knee'. Margaret rose, apparently respectfully as to a superior; 'with something of sadness in her manner, she told me you were not likely very soon to be at home'. Dorothy suggested that the whole family should move to be with their father in Edinburgh, where lodgings were cheap and 'provisions and coals not dear'. 'I do not presume to take the liberty of advising the acceptance' of any proposal, she added carefully to De Quincey, to whom Margaret had begged her to write, adding reassuringly that the little cottage and all seemed comfortable.[45] The advice was good, and the description fair, yet the letter lacks Dorothy's own affectionate tone of voice; somehow it suggests the report of a charitable visitor rather than the confidences of an old friend. Something about the De Quinceys made her constrained. De Quincey followed her advice and in 1830 moved his family to Edinburgh, where they endured a hectic existence, hiding from duns and debts. When faithful Margaret died of typhus fever in 1837, he came to brood with increasing bitterness over his feeling that she had been slighted by the Wordsworth family. This was the emotional background to his *Recollections*. Yet Dorothy and the days of their friendship he never quite forgot.

The winter of 1826–7 was a somewhat melancholy season.

11. 'in tenderness of heart': Dorothy Wordsworth in 1833

25th Aug 1842.

Recollection of Faces, my visit Rydal Mt, Wordsworth, poor Miss W.

12. 'poor Miss W.': Dorothy Wordsworth in 1842

Floods of rain in January kept Dorothy indoors, correcting proof sheets for the five-volume edition of William's poems, already in the press.[46] The weather changed in February and March to snowstorms and hard frost. Coaches were adrift in seas of snow, on which the sun glittered until 'it is painful to look out of the window'. There was 'no stirring out . . . could not go to church . . . snow too deep on Terrace . . . a dreadful night'. Dora fell ill with fever and 'a bad hoarseness', lying for a time 'at death's door'. In determinedly good heart she recovered and went out on horseback, but now there was no denying this was the 'beginning of a dangerous illness' for Dora. On 5 February their generous friend Sir George Beaumont died; Dorothy's chief consolation was that he had been 'a pious Christian . . . in constant state of preparation for the last awful change'. At the end of this year's journal she copied texts to illustrate the 'inherent depravity' and 'moral weakness' of mankind, the 'guilt that cannot by our efforts be overcome'; only 'the Saviour's grace sufficient to destroy the . . . carnal mind'.[47] As always, Dorothy responded to the spirit of her times; these were the stern doctrines which, in a parsonage across the Pennines, Aunt Branwell was teaching to the little Brontës, dogmas repeated in thousands of households throughout the land.

Spring came at the end of March 1827 with the shepherd by the Grasmere road, 'piping to his flock'. On 3 April William suddenly proposed a trip to Ulverston, the iron and ore port which had been added to his stamp district. Dorothy, at a morning's notice, set off through the rain by his side. Ulverston she found a 'Dirty ugly town', but on the homeward journey they took a boat across Coniston Water. As the rain cleared off and the mountains were gradually revealed, she fell into conversation with the rower. 'Boatman no scholar . . . hard bringing up—a parish apprentice but long served the same master—always friendly & good to him except being kept on Road driving a cart which hindered learning.'[48] It was like one of the travellers' tales by the kitchen fire at Dove Cottage, long ago. 'Saturday walked with Mary to see Jenny Fletcher and Mary Williamson—beautiful green fields—cherry blossoms—all the same as 20 years ago—all but the inhabitants.' A week later Mary went to old Mary Williamson's funeral, but Dorothy had 'caught a hoarseness. . . . I wish I could have gone also but am not well enough.' Another week passed and she was still 'not well of cold'. On 3 June she was even, exceptionally for her, 'unwell & tired . . . & not at Church'.

Relations at Halifax—the wide cousinage of her mother's family and Aunt Rawson, so indestructible that she later survived even the accident of setting her cap on fire—determined that Dorothy needed a complete holiday from domestic duties. She travelled to them on the coach from Kendal on 15 June 1827, in time for the Halifax Midsummer Fair of her childhood and two months of family pleasures in which she was the guest of honour. She was welcomed in turn to dinner-parties at each house of these solid mill-owners and managers whose fortunes had grown with the growth of the Halifax woollen trade: Mill House itself, Nab End, Stoney Royd, Ovenden, Haugh End. Everywhere young cousins came to greet her: 'the boys from Bullace Trees . . . the girls from Ash Grove'. She who 'tramped' in all weathers wrote in wonder, 'Rain comes on—carriage sent for me.'[49] There were calls with Aunt Rawson on old friends, walks in the town and gardens. She was out on the moor at half-past six on a lovely morning in July, and walked home after dinner through Birks Wood, where she and Jane had played as children. On Sundays, as well as to church, she went to the Northgate End Chapel, filled with schoolgirl memories. She saw, repelled yet fascinated, an assembly of 'Ranters . . . . groans, screams, singing "It is the Lord"'. Each day offered hospitality: 'out to tea, dinner, breakfast . . . Tea & Quadrilles . . . dinner—not home till near eleven'. She yielded gratefully to the mild worldliness of Halifax, with 'gowns made up . . . undercaps . . . Blue Ribbon', and in the New Year of 1828, on the advice of the 'fashionist' Sara, a 'fantastic . . . *black* Cap edged with blue' ordered and sent from Edinburgh.[50] Her modest accounts even reveal 'Lost at cards 2-6d.'.

The return to Rydal had its own joys. 'Lovely autumn morning—gleamy—mild—Robins singing' in September, and in October 'Rooks—Robins harmonious—loveliest of days'. On 13 November 1827, at fifty-five 'stronger than usual at our years', she made a mountain expedition with William Pearson, the antiquarian farmer of Crosthwaite: 'low clouds on hills . . . walk to prospect—mists clear & ascend Helvellyn—they gather

'—only rooks and abominable Gulphs visible—perplexity—safe descent . . . close of Day' and triumphant homecoming in the dark at half-past seven.[51] For much of early 1828 Dorothy was sole housekeeper at Rydal Mount while William, Mary, and Dora made a round of visits. Again in June she stayed behind, while William 'with our old Friend S.T.Coleridge' made an impromptu tour of Germany. Dora took the part of woman companion, which in 1798 Dorothy herself had played, but the older woman showed no signs of regret, though 'we long to have them home again'. As always, she took her home duties seriously, 'Sewing curtains all day', and Sara wrote feelingly of the 'agonies of cleaning'.[52] Dorothy's chosen pleasures were 'delightful weather for growing' and to be 'hard at work in the garden till afternoon'.

On 19 July 1828 she set out for her own holiday in the Isle of Man with Joanna and brother Harry, the 'superannuated Sailor'. She was called at half-past two and breakfasted by the kitchen fire; a walk on the misty Terrace made her 'sad at the thought of my voyage'. She wanted to go, but to leave Rydal was painful. So was the view of her own birthplace, neglected and dilapidated, from the Cockermouth coach. 'Life has gone from my Father's Court.' She stayed three weeks at Douglas, walking with Harry in the sights, the cliffs, the flowery lanes, the thatched cottages 'like waggon roofs of straw'. Ships and sailors, with their memories of brother John, stirred her imagination. 'The moon rose large and dull, like an ill-cleaned brass plate . . . and sends over the calm sea a faint bright pillar . . . boats in motion, dark masts and eloquent ropes.' By daylight the sailors revealed themselves 'Beauish tars . . . in bright blue jackets, their badges on their breasts, straw hats trimmed with blue ribands'. Yet the journal faded out, as though from weariness or preoccupation. She went home to a house full of visitors, and on 5 October 1828 wrote, 'I walk up valley of Waterfalls alone.' She had much to think about on the solitary path, for at nearly fifty-seven she was faced with new scenes and new duties.

17. 'This quiet room'

A decision so serious could only concern one of the family to whom Dorothy had dedicated her life—in this case her favourite nephew. She had agreed with Sara that it was 'a sad error' for William to despair of his eldest son,[1] and rejoiced when John was entered for his father's college at Cambridge. John, disliking mathematics, was 'strongly persuaded that the system at Oxford would suit him much better', and William, with the help of Keble,[2] the Oxford Professor of Poetry, transferred him to New College, where he matriculated in May 1823. Inevitably, according to his aunt, he suffered his share of Wordsworth ailments. Even before his first year it was a 'swelled face' caused by 'a determination of blood in the head', and in his second year, with the risk of 'Typhus in the University', his anxious parents simply kept him at home during the Lent Term.[3] In his third year William was already writing round to influential acquaintances, the Lowthers, the Beaumonts, the Bishop of Chester, to solicit a fellowship for his son,[4] but John was ruled 'ineligible'. In November 1826 he took an ordinary degree, though Dorothy quickly pointed out the Tutor's kindly opinion that he might have taken honours had he not been ill.[5]

The only alternative for John was to take holy orders, for which he would first need title to a curacy. 'Can you hear of a curacy for our John?' wrote Dorothy in November 1827, when he had been waiting a full year. Pathetically, she asked Crabb Robinson to put this young man of twenty-three up for the Athenaeum. 'It *may* be useful.'[6] Crabb Robinson, divining his old friend's daydreams, replied cheerfully that when John was a bishop he could become a member by right. Meanwhile John's disappointments contrasted with the triumphs of his Cambridge cousins, of which to his credit he was generously glad. Between them Christopher's sons won Craven Scholarships, the Porson prize, the University prize for verse in Latin, and the Chancellor's Medal for English Verse. At school they had also been notable athletes, Christopher called 'the Great' at Winchester, while Charles played in the first Eton and Harrow match.[7] When the boys—'Boys I must still call them'—arrived at Bowness for a reading party, Dorothy invited her late brother Richard's young son to visit. Her reason was the key-note of her whole life, an overruling devotion to her family; 'it pleases me to think I shall see all my six nephews together'.[8]

John's long wait ended in March 1828 when he was ordained deacon for Whitwick, a colliery village near Coleorton, in the gift of Sir George Beaumont. This was received with less than rapture by his family. 'I wish he had a *tidier* curacy,' wrote his aunt Sara, and William was frankly disappointed; 'there are not many places with fewer attractions or recommendations than Whitwick'.[9] Dorothy, though, knew better than anyone the difficulties John had overcome and was proud to see him a clergyman at last; 'John gave us much satisfaction in the pulpit and reading desk of our little Chapel,' she reported to Miss Jewsbury in Manchester.[10] John's mother set him up with household linen and supplies at his new vicarage, and the homely Westmorland housemaid, Mary, was elevated to the rank of housekeeper. Yet the industrial village was 'lonely as to society' for a young man, and John turned instinctively to his earliest protector and friend. 'Poor Fellow!' wrote Dorothy indulgently; 'he wanted me sadly to go with him.' It was a foregone conclusion; she could not resist this appeal for sympathy. She promised to go to Whitwick in winter as John's fire-side companion. All other plans were set aside, including hopes of an Italian tour with Crabb Robinson. 'Alas for Rome!'—she could not see it 'even in a day dream'. She asked for letters 'to cheer my solitude at Whitwick . . . with John in his lonely Parsonage'.[11] On 7 November 1828 she set off by the coach from Kendal. For the next week she stayed with the Jewsburys in Manchester, evidently preparing for her duties in an industrial parish. 'At Cotton Factory . . . but first at Infant School,' she noted, 'shirts dirty . . . at Institution, Dirt & Dust Institution & Thread Factory.' There was no rain, but a heavy grey gloom over the streets.[12] Yet she remained cheerful and Maria Jewsbury noticed her genius for making friends. 'She won all hearts before and around her . . . an embodied spell.'[13] Bravely, Dorothy climbed the ladder to the top of the coach for Nottingham and was rewarded in spite of winter with 'brightness all day . . . Darley Dale enchanting—Matlock the like', and the sight of Haddon Hall clinging to the slopes above the River Derwent. After a night at Nottingham she set off by eight o'clock, with 'sun risen an orb of gold', for a day's driving over clear roads. The same evening, 18 November, she was at Whitwick; 'met John—much pleased to find myself within his lowly parsonage'. Within a few days she knew herself so welcome to John in his loneliness that she was determined not to leave him to winter solitude. 'I of course will remain with him.'[14] She spent the first week arranging, making curtains, mending John's old coats. In the evenings John read the newspapers, while she sewed, and read over his yesterday's sermons, or Isaac Barrow's tracts.

In some ways it was curiously like the girlhood dream of 1793, at Forncett, of life with William in 'the Happiness . . . in my little Parsonage'.[15] She was mistress of the thatched cottage ornée, the curate's trusted critic, drafter of accounts for charity as in the old Forncett days, and 'more useful than I could be anywhere else'.[16] John was beloved as William had been, with the same all-absorbing, self-effacing devotion. Yet just as John lacked his father's genius, Dorothy could no longer draw on the boundless energy of her young days. She was elderly, tired, sometimes ailing. As for Whitwick, 'never was there a place more barren of society'.[17] The three villages of the parish were 'in a sharp and cold situation' in the hills, cut off by rocks to the north and Charnwood Forest to the south and east.* There were none of the peasant freeholders whom Dorothy loved and admired; the meadows and commons were enclosed, and the inhabitants worked in industry: a brick-kiln, several stocking-mills employing women and children as frame-work knitters, while the men worked overwhelmingly at 'digging for coals'. The landscape was pitted with soughs and drains for carrying water from the mines and scarred by engine-houses for pumping; more than forty horses and carts hauled coal over the single high road.[18] The people were a tough and forceful breed of dissenting stock. 'Our warfare with the Methodists' was not a vicarage joke; Dorothy found Whitwick full of 'conventicles poverty and all the bad habits' of a crowded manufacturing village. She hoped to help John 'ameliorate the condition of the Poor', but did not expect this to be easy. Determinedly cheerful, she wrote to Crabb Robinson of 'the pleasure it will be to

---

* The area is now covered by open-cast mining.

me during the coming lonely winter to receive tidings of distant
Friends'.[19] She had, in fact, friends only two or three miles
away at Coleorton, the wife of John's rector and the widowed
Lady Beaumont at the Hall, who sent kindly notes with gifts of
game from the estate. The walk to Coleorton was 'not too long
for a winter's morning call', but as Dorothy's journal shows
she was fully occupied at home.

The day began while it was still dark, with family prayers
before breakfast. Day dawned on the alien landscape without the
delights of home: 'a gloomy morning—slight rain . . . a blus-
tering dark morning . . . dark & damp'. The new year of 1829
brought 'foggy black frost . . . very cold—sharp wind—rheumatism in the morning—stiff joints & sore flesh' for the first time
in her life. John's congregations were 'poor . . . wretched congregation in the morning', but his aunt blamed the weather.
Morning continued with serious reading: Greek for the curate
and for Dorothy strictly devotional books. To help John each
week she 'read John's sermon—very good'. The parsonage did
not seem to offer novels, plays, or verse, other than William's
poems on religious subjects; 'W's sonnet very appropriate'
against 'prejudice, intolerance—hatred to the Church'.

Housekeeping in the thatched cottage was modest. Mary,
the servant from Rydal, praised as a 'frugal housekeeper', provided black puddings for dinner and rabbit pie on Christmas
Day, Dorothy's fifty-seventh birthday. House-cleaning, in the
foggy air, laden with coal-dust from the mines, was endless and,
by Dorothy's standards, 'Unprofitable—working at chairs all
the morning . . . intended to go to Coleorton but wrote till too
late . . . tired & could not continue my labours'. Washing day
came round more often than at Rydal and characteristically
dawned 'dark & damp' or with 'slight rain'. Soot and smuts
were the appointed enemy of clean shirts and sheets. When the
housework was finished there were sermons to copy from old
bound volumes for John to preach on special occasions.

Dorothy's chosen parish work, as in the distant Forncett
days, was to visit households in distress. In a working population of over two thousand in the townships around Whitwick
there were troubles enough, and she set off along muddy tracks
which led to straggling rows of cottages or on the highway
avoiding the coal-carts. 'Walked round by Swannington village

—Large Methodist Chapel.' She had always cared deeply for the very poor, like the beggars who called at Dove Cottage, entering into each hard life with total sympathy. This indeed was the secret of the 'spell' Maria Jewsbury had noticed. Yet at Whitwick, as perhaps the only educated woman in the parish, she found herself an outsider. Villagers observed the strange small gaunt figure, who tramped the roads in all weathers. 'Walked two hours on Leicester road and among rocks—was asked if I had lost myself by ragged Girl nursing Baby.' Even after some months sharp eyes observed her: 'walked . . . cross Mrs. Howe milking—tells us we are trespassing'. She did not give up and by March 1829 there were homes where she was welcome. 'Only short walk—called on various poor people.' From her own small means, she gave what she could. Her accounts record items such as 'Charity woman 1-6, Poor Woman 6^d, Boy going to America 1/Old Anne West 2/ Arrowroot & Sheepshead 9^d, Billy 6^d for task, Sick woman 5/6'. She 'called on poor woman very ill—oranges . . . John obliged to go to pray with poor woman . . . after dinner at Billy's Child very ill . . . Wife told me the story of her lying in—large appetite & nothing but Tea & dry Bread'. One hard-pressed wife even told the story of her early years in 'servitude with old Mrs. M.'. Unexpectedly they were the happiest in her life, for the rather pathetic reason, 'so loving & friendly all together'. These visits brought Dorothy satisfaction, for she felt she was helping John. 'I foresee that I shall not find it an easy task to leave him *finally*, in a place where I feel myself of so much use to him,' she wrote.[20]

She did accept one invitation, though. Whitwick was within a day's journey of Cambridge and on 6 February 1829 she went to spend a month with Christopher—now respectfully called 'the Dr.'—and his sons. The young men devoted themselves to her, with walks in the college gardens, three-mile tramps by the river, a supper in John's rooms—'Oysters & c'. —and a visit to King's Chapel—'sermon in bad taste'. Dorothy loved them, and was grateful, but her journal shows how strictly she now confined herself to religious duties. There is not a word of her old delight in the ancient buildings, the Fitzwilliam pictures, the college choirs. Instead she kept a record, with criticism, of each sermon she heard; 'Universality of the Gospel at St. Mary's . . . voice weak & not always audible'. The

Round Church was slightly more satisfactory, though 'pressing his point too far & the arguments not clear enough'. Her brother gave several dinner-parties at the Lodge; she met on equal terms as a guest Adam Sedgwick, first Professor of Geology, William Whewell the natural scientist, Connop Thirlwall who opposed her brother on compulsory religious teaching, and Lee, the Professor of Arabic and Hebrew. Yet Dorothy said nothing of these distinguished intellectuals or their conversation, apart from 'large party at dinner' or 'pleasant evening'. Instead at the dinner on 4 March, the final evening of her visit, she devoted herself to the one guest with a tale of family troubles to tell: 'Wife's sister buried 3 children in the hooping cough at Boulogne & 3 months after husband in a consumption.' Next morning, Dorothy left Cambridge at seven o'clock.[21] She would dearly have loved to see Catherine Clarkson. 'What should prevent you?' asked Henry Crabb Robinson, with the nearest to impatience he ever showed. 'It is only a day's journey on the Ipswich coach.'[22] Yet apparently conscience would not allow her any further holiday; she hurried back to Whitwick, to a bitterly cold month and a series of domestic crises.

All through March the wind blew incessantly from the north-east bringing frost, snow, sleet and hail or heavy rain. Mary the housekeeper had a very bad cold and there had been some trouble with her and other helpers; Dorothy wrote a letter to Rydal '& destroy my letter', in order not to worry the family. Half-way through the month, she and John walked to Coleorton Hall in the teeth of the wind. She was so chilled that Lady Beaumont and her sister put her to bed and sent her home by carriage next day. 'A bad hoarseness' kept her in bed on Sunday—an unprecedented indulgence—but she was up on Tuesday, 17 March for washing-day. A week later it was 'John not well'; he could not take duty at the church or attend a 'Protestant meeting' with his rector. His attack of influenza lasted a week, during which Dorothy sat with him, reading the Bible aloud and mending stockings, and at the end of the month she worried that he was 'still very poorly'. She struggled on, paying bills, settling accounts, visiting sick 'poor people' in the influenza epidemic.[23] Cold, fatigue, and distress had always been enemies of her health, but determinedly she fought them off.

On Wednesday, 3 April 1829 she suddenly collapsed, with paroxysms of pain in the intestines. For 'forty-eight hours she was in excruciating torture,' wrote William, calling it 'internal inflammation'.[24] Pain so severe, and prostration, have suggested a modern diagnosis of gallstone and acute inflammation of the gall-bladder with biliary colic.* Certainly this attack was sharper than any she had known: 'Sickness . . . hot and cold —with pains in Bowels, daily and constant companions.' She heard the old Whitwick women at her bedside talking to each other 'as if quite sure I *must* die'. She was reduced to such weakness that she could not stand and could hardly speak; she whispered that 'She was alive—as her poor grandfather used to say—and that was all'. 'Were she to depart', wrote William in horror, 'the Phasis of my Moon would be robbed of light to a degree that I have not courage to think of.' In this crisis, on 10 April, Mary Wordsworth arrived to nurse her. After a week she was pronounced out of danger and on 19 April—with infinite gratitude—was 'out in the garden for ten minutes'.[25] In May, she took stock of her situation. John had been offered a living as a 'gratifying' favour by Lord Lonsdale and she hoped he would marry, tacitly admitting that her own long service to him must come to an end. She had never complained of the hardships of Whitwick, but at the beginning of June she again felt 'not well'. Instantly, Lady Beaumont's sister fetched her to the Hall, and there was an understood admission of hardship in Dorothy's gratitude. 'Dear Coleorton all elegance and chearfulness at our return.' They were warned she might suffer a relapse if excited or tired,[26] and kept her very quiet for a week. She was well enough to welcome the new curate, attend John's farewell sermon in the afternoon of Sunday, 21 June, and pack 'my own clothes'. On Monday Lady Beaumont came to dine— '& on this day I parted from her'. It was the end of an epoch.

Dorothy's dangerous illness had 'left a homesickness behind it', but she could not travel north without visiting Aunt Rawson of Halifax, widowed the previous year. She found the old lady of eighty-three with all her faculties, except for slight deafness, 'absolutely healthy, so tranquil yet so chearful' in her widowhood that Dorothy hoped for many more meetings in this

---

* See Appendix Two.

world.[27] Martha Ferguson—'Patty' of their shared schooldays
—seemed 'very feeble' and Ann, though healthy, aged in voice.
Cousin Edward proudly displayed the elegance of his cottage,
Bullace Trees. Family carriages called to take Dorothy to the
old church, long rounds of morning calls on friends, to the
Library, the Gardens, or gathering roses at Hope. 'Young
folks' took her for an evening walk by the canal, or acted
charades. It was delightful, but, as she wrote to remind herself
on 11 July, 'all nothing save Christ Jesus'.[28]

As though to confirm the transitoriness of life, she had a
relapse into pain and fever in the first week of August, traced
day by day in her journal: 'not well . . . very ill . . . weak & ill . . .
middling . . . almost well . . . very much better . . . out in
garden . . . quite well'. The illness was vaguely diagnosed as
cholera morbus—not the lethal Asiatic cholera, which did not
reach Britain until autumn 1831, but the old term for any
enteritis with pain and purging. Dorothy found it 'not so
severe' as her last attack,[29] and recovered to escort her old aunt
on a visit to Mr Rawson's sister at Saltmarsh near Goole. She
expected to find the flat country ugly, but even illness could not
destroy her gift of perception. Her old pleasure revived to find
'the River Ouse passes the door . . . constantly changing with
sloops, boats, ships etc. etc.'. On fine evenings she saw 'at
Sunset golden Ships reflected in the broad full River'. Aunt
and niece returned to Halifax, tired yet happy, for a last family
evening with all the Fergusons to dinner on 6 September. Next
day, after 'an affecting parting from our dear old Aunt', she
took coach for Bradford and Kendal, then pony-chaise to Rydal,
'thankful and happy to find myself at home again', after a narrow escape from death.

Yet when Dorothy arrived at Rydal Mount in the second
week of September 1829, she seemed at once to realize that life
would never again be the same for her. On 14 September, she
walked for only two miles, resting at the end of one mile and
yet still coming home 'fatigued—and worn', and needing an
hour's further rest.[30] For the first time, she admitted age. 'We
old ones', she wrote on 27 September, 'begin to look a little
aged.'[31] Returning from a visit to Ireland early in October,
William thought her 'very unwell', and she herself, while
recording small improvements, confessed, 'there is a sense of

weakness with unconquerable stiffness—I say unconquerable', she added, 'because it goes away so slowly'.[32] She could still mock at her invalidism—'A whole page all about myself!'—but it was real, and, in the pages of the journal, which she still kept, the entries grew temporarily confused and almost illegible. 'She does not', commented William,[33] 'throw off the effects of her Whitwick illness', and she herself confessed on 19 November that 'exercise amounting to the slightest degree of fatigue invariably disorders me'. She was delighted with 'the lower new-made green terrace' in the garden, which William had constructed in her absence, and allowed herself, on her less good days, to be dragged around the garden in a bath-chair by James Dixon the gardener, whose bow was 'the most exquisite sample of respectful Simplicity I ever saw'. Yet she was far from strong as she approached her fifty-eighth birthday. 'Of one thing you may be secure,' she wrote to an old friend of William, 'that you will never now find the house emptied of its inhabitants; for I am always at home.'[34] This proved more true than she knew.

In these years of illness, Dorothy was attended by Thomas Carr of Ambleside. He was a far cry from their old apothecary, of whom she had sometimes sought a first opinion when he was drunk and a second when sober. Carr was in his early thirties, and by the standards of the time well qualified, with five years' apprenticeship to a Lancaster surgeon, lectures on anatomy and physiology, theory and practice of medicine, and longer than compulsory clinical attendance at the Royal Infirmary, Edinburgh, after which he was registered as LSA, MRCS.[35] Although, as we have seen, a differential diagnosis of gall-bladder inflammation has since been made, Carr, who attended her for many years, made the first professional attempt to investigate 'Miss W's bilious attacks'. The term 'bilious' was a nineteenth-century euphemism for bowel disorders.[36] Dorothy's spasms of acute pain 'in the Bowels', the violent vomiting and purging, the relapses growing more frequent with the years, led him to diagnose in February 1830, 'the disorder was inflammation of the Bowels'.[37] This condition, now usually known as spastic colon, became under the term 'mucous colitis' a mainstay of fashionable nineteenth-century spas. Dorothy complained of pain in the bowels, never in the side or back. The onset was usually in early middle life; Dorothy's first recorded attack, when she took laudanum and 'lay in bed all day', was in October 1801, two months short of her thirtieth birthday.[38] The subjects were most often women of emotional temperament, vulnerable to excitement, worry, or stress, and given to sudden weeping or laughter. Dorothy herself wrote at this first illness, 'I eased my heart by weeping', and Mary noticed how often at family crises, like William's wedding, she was 'ill in her usual way'. The characteristic loss of flesh was noted by the Halifax cousins in 1805 and De Quincey in 1807. The harsh climate of Whitwick was a likely factor in the serious attack of 1829; 'cold is my horror; it flies instantly to my bowels.'[39]

For the next five years Dorothy's condition established a pattern: repeated attacks of illness, generally increasing in length and severity, each leaving her more disabled, with dwindling periods of remission, and, as William wrote, 'every relapse finding her weaker'. Her fifty-ninth year began ominously with a month's silence in her journal entries—'No memorandums since before Xmas.' To old friends such as Mary Lamb[40] she could write, '*Wishes* I do now and then indulge of at least *re*-visiting Switzerland, and again crossing the Alps, and stretching on to Rome', but she recognized their unreality. 'I entertain them as an amusement perhaps for a short while.' In the same vein she allowed herself to hope that 'in the Spring, I shall be able, to a degree, to resume my old habits', but when the spring came, though 'it is no punishment to be confined to this beautiful spot', she realized, 'I have been enacting the Invalid ever since the month of November', and would have to go on doing so.[41] Her delight was yet another green grass garden terrace William had constructed for her, and her joy overflowed on May Day 1830 in her journal.[42]

On new Terrace—Sun bursts out before setting, unearthly and brilliant—calls to mind the change to another world—Every leaf a golden lamp—every twig bedropped with a diamond.

She could not help adding, with a haunting sense of mortality, 'The Splendour departs as rapidly.' In point of fact, the summer of 1830 did bring a longer remission, during which, as a change of scene from Rydal Mount, she was able to make a short stay now, and another later in the year, with her cousin and namesake Dorothy Benson (née Wordsworth), known as 'middle Dorothy' (between herself and Dora), at her home at Green Bank, Ambleside. Yet she was still not well enough to be left alone for long, and Sara Hutchinson was her constant companion when other members of the family happened to be away. She kept her interest in young people, knowing of Willy's failure to find any job and his hopeless secret attempts to enter the army. Visits from Quillinan's 'dear little girls' made 'a welcome stir in our quiet habitation'. 'I am so fond of neptune the big dog,' wrote Jemima Quillinan innocently in 1831, 'he doesn't bite anybody but beggars.' Dorothy could still manage short walks, and kept in touch by letter with old friends. Early in July 1830, she wrote two letters to Annette and Caroline in Paris.[43]

She had, indeed, written another letter to France at this time, but this was probably part of a troubling correspondence with her old Lakeland friend Miss Barker, whom she had last seen at Boulogne in 1820. To Dorothy's agitation, Miss Barker announced that she was marrying a young man called Slade Smith. Dorothy, intensely disapproving, decided that he must be 'a Boulogne swindler',[44] and 'wrote my mind . . . perhaps not very palateable'. Not surprisingly, there was for some time no answer. Though Mrs Slade Smith eventually wrote promising a letter 'from her own Chateau',[45] this stout walking-companion of Dorothy's healthier days now dropped out of her life for good. At the same time, she made new friends and even experienced new pleasures nearby. Her nephew John, now settled in a Cumbrian parish at Moresby, had become engaged to Isabella Curwen, daughter of a rich mine-owner at Workington, who also had a mansion on Belle Isle on Lake Windermere. The wedding took place at Workington on 11 October. Dorothy was forced to stay at home, and her journals on that day record only a note of mild resignation—'Robins cheery—with sadness' —but she took pleasure in the wedding of William's offspring, and recorded its anniversary in later pages of the journal.

Dorothy spent her fifty-ninth birthday on Christmas Day 1830 quietly at home, having 'a small party' with Sara Hutchinson.[46] William, Mary, and Dora were at Trinity College, Cambridge, with Christopher. Dorothy would have liked to have gone with them, but, still with courageous hopes of a final recovery, 'am quite satisfied that it is, for this one winter more, the safest plan to stay quiet at home'. She read 'by our quiet fireside' books of devotion, Wesley and Paley, and calm entertainment, such as Cowper's letters. With her few visitors she played chess. Looking out over the garden in February, after a snowy and frosty start to the New Year, she noted 'snow drops in warm places hanging their bead like heads', but did not often venture out herself, being 'driven in from the Terrace by sharp snow & hail showers' when she tried later that month.[47] Summer, when it came, was exceptionally fine, and this even encouraged her to spend the first days of September staying with the Curwens on Belle Isle, an almost unheard-of adventure for her, though she did not like the Curwens' domed mansion, resembling a giant's pepper-pot, so well as familiar Rydal Mount, nor the sensation of living on an island.[48] 'What I like least in an island as a residence is the being separated from men, cattle, cottages, and the goings-on of rural life.' She was obviously optimistic, though William, seeing her early in summer after a six-months' absence, was gloomy. 'She will never I fear', he wrote to his old walking-companion Robert Jones, 'fully recovery.'

His gloom was justified. On the eve of her sixtieth birthday, at Christmas 1831, Dorothy suffered a severe return of illness. She was confined at least ten weeks to her room, and continued to be prostrated by numerous relapses. She did not write a word in her journal for ten months. In May, she received the sacrament in bed, in the company of Mary and William, in preparation for death, and it was not until August that she improved enough to be driven out in a closed carriage, belonging to a neighbour and more 'easy' than the Wordsworths' chaise.[49] Even then, reported Sara Hutchinson, they could never leave her alone for an hour. She was, however, William wrote to Crabb Robinson, 'in a contented and happy state of mind',[50] and, as he wrote to Robert Jones, 'Her mind is rich in knowledge, and pleasing remembrances; in which, and her religious faith, she finds thro' the blessing of God, abundant consolation.'[51] She even composed halting verses of a religious nature.

Such gifts are mine, then why deplore
The feeble body's slow decay,

A warning mercifully sent
To fix my hope upon a surer stay?

And may I learn those precious gifts
Rightly to prize, and try their soothing power
All fickle murmuring thoughts repress
And fit my fluttering heart for the last hour.

She faced her painful and distressing illness with the greatest courage, welcoming the family to her bedroom fireside, allowing Dora to wash her face 'as she used to do mine 30 years ago,' wrote Dora. Only once she admitted her almost painful longing to go out of doors. Otherwise she contented herself with the views from her bedroom, south over the park of Rydal Hall and west towards the setting sun.

She took up her journal again in October 1832, after ten months' silence, in a mood of resignation and deepening piety. 'My sentiments', she wrote on 4 October, 'have undergone a great change', and she instances as an example of this 'the absolute necessity of keeping the Sabbath by a regular attendance at Church'.[52] Her old sparks of temperament were subdued to a quiet humility. When her oldest friend Jane Marshall offered her an annual gift of money, she wrote, in accepting, 'so far from being ashamed of receiving such a Gift from *such* Friends I am proud of it', and planned to spend it on 'gratifications, such as I might otherwise have scrupled to indulge myself with'.[53] The independence in money matters, which she had proudly maintained for so long, was abandoned. Very occasionally, there was a flicker of her former self. On Boxing Day, 'the Fiddlers went their round. Found me awake.' Her mind, presumably, went back to earlier times, when she had danced youthfully, and, as the Coleridges had thought, unsuitably, with the children on similar occasions. It was only a temporary flash of memory. Within a day or two, she was once again seriously ill. After a day on which she had ventured to leave her room, she 'had an unusually bad night', which heralded an even more alarming attack. 'Our dear Sister is very poorly and seems to grow weaker every day,' William wrote to their brother Christopher, on 29 January 1833. Earlier Dorothy had directed him to write that she was 'very much better' and even added in her own hand several lines of good wishes to

Christopher and his son John.[54] Now she was worse than ever, with fresh and alarming developments, swollen legs accompanied by black spots, which faded when William faithfully rubbed her ankles to restore circulation. The doctor, 'apprehending mortification', considerably increased her usual dosage of opium and brandy, a palliative,[55] maintained for at least two years. 'Without the help of stimulants', wrote Sara to Mrs Coleridge, 'you would scarcely believe her alive—but with these she is kept tolerably easy—& even comfortable & cheerful.' To relieve the spasms of pain she took fifteen to twenty-five drops of laudanum twice a day,[56] and became inevitably, though innocently, habituated to the drug, which she had condemned so severely in Coleridge and De Quincey.

By mid-February 1833, Dorothy was believed to be on the point of death, and, according to her own hazy recollections of this time, 'my poor Brother went to lie down on his bed thinking he could not bear to see me die'.[57] Looking back nearer the time, she could hardly remember anything that happened. 'I would fain make a record of the past but alas! it is almost a blank'; she could only recall being 'very helpless as to reading and working'.[58] It appears she could still make occasional entries in her diary which spoke with her true voice. At the end of April

Two glowing anemones and a snow white companion are in a pot on my window ledge, and two knots of primroses of the Alpine purple. Rooks busy all the birds of the sky and earth are singing and all is wrapped in happy brightness.[59]

From these six blank months in 1833, she never fully recovered. Although her reading habits returned, and she was grateful for Crabb Robinson's gift of Gilbert White's *The Natural History of Selborne*, a book she had always wanted to possess, he found that in conversation she was 'too nervous for disputation'. When in autumn 1833, a self-taught local artist, Samuel Crosthwaite, came to paint a portrait of William, he stayed to paint Dorothy, a likeness 'begun for charity—but as the work proceeded our charity turned into gratitude to the little man for putting us into possession of a thing so valuable'. The value of this simple portrait was, of course, enhanced for Dorothy's family by the feeling they had so nearly lost her for ever in the early part of the year.[60] She sits, a tiny shawl-wrapped figure in a lace cap, upright in an armchair, with writing-case on her knee, pen and ink beside her, and a pair of spectacles in one hand.[61] Watchful near her is the pet dog, Miss Belle, who shared her room, and beyond her a window gives a view of the Rydal Mount garden and the Lake scenery which had, she felt, such a healing effect on her convalescence. William, who had read the recently published *Later Essays of Elia* to her, while at his task of rubbing her legs, reported to Lamb earlier in the summer, 'I have been this particular, knowing how much you and your sister value this excellent person, who in tenderness of heart I do not honestly believe was ever exceeded by any of God's Creatures.' The portrait, much loved by the family, perhaps embodies this view, though her unrevealing, calm expression may owe something to the doctor's use of opiates.

Even with a remission during this autumn and winter, there was, she wrote, 'no rapid return of thought', though 'I am much stronger in limbs and can walk without stick or arm'. What thought there was turned more and more to religion; during the winter months of 1833-4 she read her Bible daily, and had 'finished my 3rd reading of the New Testament' by Easter 1834.[62] Earlier in the year, she praised 'a conclusive article against Liturgical Reform' in a magazine. 'Crossing' some of her journal entries about this time is a short single verse.

> A prisoner in this quiet room
> Nature's best gifts are mine
> Friends—books—and rural sights and sounds
> Why should I then repine.

She told Lady Beaumont, 'my prison (if we may so call it) is one of the prettiest and most chearful in England'.[63] Nor was she oblivious to the darker doings of the outside world. That same early and bitterly cold Easter 1834, she wrote of

A poor woman refused a lodging by Thomas Troughton on Tuesday night slept with her husband in Smithy hut without straw or covering —next day proceeded & was delivered, by herself, of a dead child on road near Quarry.

Her feelings for all forms of life were heightened by her religious sympathies. On Sunday, 6 April 1834, there was

Loveliest of Sabbath mornings. How I long to be free of the open air!
The Birds are all singing. Rooks very Busy. The earth & air & all that
I behold seems a preparation for worship & sabbath rest. Bird & Beast
have this one days security, from snare or slaughter.

A week later 'with a thankful heart I revisited the garden and
green-terrace in my little carriage'.[64]

On 27 July 1834, Dorothy entered in her diary a laconic note
of the end to a long chapter in her life.[65]

This morning came the sad tidings of poor Coleridge's departure
from this world. He died in a pious & happy state of mind & had
many hours of ease before his dissolution, after great suffering in the
Bowels.

Did she remember anything of the moment, thirty-seven years
ago, when he had burst into her life, leaping the gate and running across the Racedown field? She may have had her memories, for she had 'a long sit with knitting on my knee'. Would
he himself have recognized in this pious invalid the 'perfect
electrometer' of taste he had celebrated then? In his old age,
Coleridge became very dismissive of the qualities of intel-
ligence in women. In 1830, he laid it down that it 'was the
perfection of women to be characterless', praising them as
'creatures who, though they may not always understand you,
do always feel you, and feel with you'.[66] Less than a year
before his death, he had actually used a pointless and perhaps
untrue anecdote about Dorothy to support his contention that
one of the chief charms in women was that they did not understand a masculine sense of humour.[67] It was perhaps as well
that Dorothy and Coleridge, in their sixties, never met, and
that she could have a vision of happier days as she sat dreaming
with her knitting on her knee.

She certainly had happier times this summer. The two Hutchinson boys, Tom and George, whom she had noticed six years
before at Brinsop as so well brought up, came to stay. They
dragged her bath-chair round the garden for an hour at a
time,[68] and read to her indoors. She specially liked Tom's
reading, 'as if he understood & *like a Gentleman* which Geo. does
not'.[69] She even attempted a very short walk herself, 'once
round the Gravel Front of the House'.[70] She wrote to two of the
boys' sisters, at Brinsop, Elizabeth and Sara Hutchinson. To

the former, her fourteen-year-old god-daughter, she wrote:

I send you a Godmother's Blessing, with sincerest wishes that you may not waste the happy days of Youth. Make the most of them. They will never return, and if you do not profit by present advantages you will bitterly repent when it is too late, but however happy you may be in the enjoyment of youth, health and strength never, my good child, forget that our *home* is not here and prepare yourself for what will come, sooner or later to every one of us.[71]

It sounds like her own voice, speaking to herself, particularly when she added 'How I wish I could ever again climb the Credenhill hill with all of you—young and old!'

Autumn 1834 found her handwriting noticeably deteriorating, though 'my excuse is, that I lie upon my back in bed and with uplifted knees form a desk for my paper'. In winter and New Year she had a series of the attacks of pain and purging and lost ground. During this winter of 1834-5, Dorothy had long periods of weakness, followed, in the now familiar pattern, by renewals of strength. 'Of late have been well and ill.' Yet she enjoyed company, and mild, useful activities. She was visited, on 6 October 1834, by Diana Dixon, apparently a village sempstress. 'She is a good-natured obliging Girl whom I always like to see. She took those 2 Ruggs to fringe for me.' Even when mainly confined to bed, Dorothy could not be idle, and in spite of cramped position and stiff hands continued to braid rugs all this winter. Yet her activities were often interrupted by bouts of acute pain—'30 November, weak with pain', and the next day 'Weak & in pain'. From 7 December 1834 to Christmas Eve, the day before her sixty-third birthday, she was in bed, too weak to keep her journal, and suffering pain which she described, when she resumed writing, as 'piping agony'. She happily used the word in another sense when she wrote of this period, 'my own companion Robin cheared my Bed-room with its slender subdued piping'. She could even at this time compose verses of self-comforting.

> My failing strength my lolling limbs
> Into a prison change this room
> Yet is it not a cheerless place
> A cell of sorrow or of gloom.

Into this quiet room, early in April 1835, came a reminder to Dorothy that the past too was still with her.

In the later part of 1834, William was agitated about expenses. Though he had made a number of investments, in both English and American bonds, he feared that his earned income might diminish and that there might be additional calls upon it. Reorganization in the Distributorship of Stamps threatened to lose him payments, and furthermore end Willy's job as his father's sub-distributor, which he had held since 1832. William therefore sought to substitute for the unofficial annuity of £30 he sent to his daughter Caroline Baudouin a lump sum to be invested in French bonds by Caroline and her husband. He put this matter in the hands of Crabb Robinson. Unhappily, the £400 William felt able to authorize was far from enough capital to maintain the rate of the annuity. Instead of the £30, which the Baudouins had received annually since 1815, they would now, after investment, receive barely half that amount. It is clear each party, William and Robinson on one side, the Baudouins on the other, regarded William's payments differently. On the English side, they had been reckoned as *ex gratia* payments, and capable of alteration. On the French, they were regarded, naturally, under French law, as the parent's obligatory contribution to the marriage portion of his child.[72] The whole situation made for irreconcilable misunderstanding, which prolonged negotiations all through the first quarter of 1835, and led Robinson to enter into his diary the unjust opinion that 'the Baudouins are trying to extort money without any good feeling or excuse whatsoever'. On the French side, Caroline's mother, Annette, was agitated and bewildered. She was sixty-eight now, and nearing the end of her hard life. In despair, she wrote again, at the beginning of April 1835, this time to Dorothy, addressing her at what she called 'Grasner' (Grasmere) and 'Ridelmonde' (Rydal Mount).[73]

She wrote to her 'dearest Dorothy' that she was still awaiting a letter from William with a prompt fulfilment of her 'just claims'. She understood that his own family circumstances might burden him, but surely he ought to carry out his firm promise. What he now offered bore no relation to the sum in the marriage contract, and Baudouin as a good husband and father was bound to demand his family rights. Remembering

William's scrupulous honour and the nobility of heart she had always admired, surely the thought of this debt must distress him?

As the letter went on, Annette's feelings, so long restrained, broke out with all the pain of a mother who feels her daughter unjustly neglected. 'My dear daughter, so admirable in the qualities she shares with her father, deserves all his tenderness because her standards of good and evil resemble this. She asks me to offer him all her respect and affection and embraces him with all the warmth of her heart.' Caroline's little girls, too, do not deserve to be ignored. 'Your nieces are both charming they are beautifully brought up and the younger will soon make her first communion.' Briefly Annette recalled her own past with its painful memories and her present griefs, which should cry out in William's heart. 'And you, kind Dorothy, whom she loves and trusts so much . . . take your niece's part and let her mother die in peace—I care little for life and god may cut me off whenever he pleases.' She ended, 'Adieu, your unhappy Annette W.'

This heart-breaking letter was folded small enough to tuck into the back cover of Dorothy's last journal, where it survived hidden apparently until the 1970s, perhaps escaping family censorship and destruction. It was the last word from Annette who, bleakly described as an employee aged seventy-five, 'known as Williams', died in 1841. This letter with defective grammar, phonetic spelling, phrases wrenched from the depths of feeling is touching in its simplicity. Unhappily, it is impossible to say how well Dorothy understood its appeal. The letter reached England on 4 April 1835, but by 3 April Dorothy's own journal handwriting had become wavering, almost illegible. It was the onset of her long-drawn decline, which now took a new form.

18. *'Oftener merry than sad'*

Dorothy seems to have recognized the onset of her own mental confusion some time in the year 1835. Though they cannot exactly be dated, three lines of verse in her last journal suggest her self-awareness:

> My tremulous prayers feeble hands
> Refuse to labour with the mind
> And *that* too oft is misty dark & blind.

The first clear sign of imbalance occurred on 17 February 1835; it took the form, common in early senility, of irritation with those nearest to her. For the first time, there is open criticism of William and Mary. Early in February they planned a long visit to London, to lobby the Prime Minister, Peel, to use his influence in finding a safe job for Willy. Peel, whose Government was threatened, had declared his inability to help on 3 February, only to receive a lengthy letter of appeal from William, written on the 5th.[1] A fortnight later, William followed it by a visit with Mary to London, to plead his son's cause. He had evidently discussed this with Dorothy who, normally, would agree to anything which might help her nephew. Now, faced with what she felt desertion of herself and Dora—though Sara remained to care for them at Rydal—she burst out:

Wm & Mary left us to go to London. Both in good spirits till the last parting came—when I was overcome. My spirits much depressed. . . . More than I have done I cannot do therefore shall only state my sorrow that our Friendship is so little prized & that they can so easily part from the helpless invalids.[2]

Her distress continued in uncharacteristic, almost childish, entries. On 21 February, 'No letter from Wm & M. to me a great disappointment' and two days later, 'W. & M. too full of business'. Her writing deteriorates, and by the middle of March is nearly illegible, though she was still capable of following the news: on Friday, 10 April, 'All public affairs chearless! This morning hear that Mr. Peel is out of office!' Since, as William said,[3] 'The business which brought me to London arose out of a hope of procuring some respectable situation for my son Wm

from Sir R^t. Peel's Government', there was no more reason for
him to remain away, though he lingered till 24 April, staying
with Christopher at Cambridge: By then, Dorothy had written
and misdated (9th for 19th) virtually her last journal entry, an
incoherent page beginning 'Another week & nothing set
down'.

Her new state, apparently fluctuating, was puzzling to those
watching over her. On 4 May William found her 'rather better', but on 13 May 'much weaker'. On the other hand, about
18 May, she had a highly lucid interval, and wrote a letter to
her nephew Christopher,[4] which was both legible and affectionate. She was thanking him for the gift, 'from your apartment in Trinity College to my quiet prison-house', of a print of
the Virgin and two babes, which 'when all the Family are gone
to Rest is my soothing companion when lighted up by the tem-
perate blaze of the fire'. This was possibly a reproduction of
a Raphael, which Christopher could have seen in the Louvre
during his recent European tour, the so-called 'Belle Jardinière'.

At the end of this month came an unexpected reversal of
fate. Sara Hutchinson, the family sheet-anchor, fell ill with
acute rheumatic fever. Though, for a time, delirious, she
seemed to be on the way to recovery, and anxiety was still concentrated on Dorothy. 'My Beloved Sister's days are drawing
steadily to a close,' William announced to Southey on 7 June.[5]
Yet it was the unexpected that took place. On 23 June 1835
Sara had a cerebral haemorrhage, and died almost instantly
with 'no acute suffering'.[6] Resigned to Dorothy's coming
death, all were shattered by Sara's. The cheerful, lively, bless-
edly common-sense housemate, survivor of the darkest days
with Coleridge, tease of William's glooms, was gone from them
for ever. As for Dorothy, William brought himself to believe
that 'her mind received a shock upon the death of Miss Hutchinson from which it has never recovered',[7] though she did
not apparently speak of her old companion, and in moments of
realism, he admitted that signs of mental confusion had probably appeared before this event.[8] Sara and Dorothy kept their
place in the family circle as 'Our two beloved sisters, the one
gone before and the other changed'.

Dorothy herself seems, obscurely, to have felt she had
become too dependant on the large dosage of opium Mr Carr

was giving her, and which she took twice every twenty-four hours. For the first time she voluntarily suggested she should not take it, and 'although she had moaned a good deal' seemed no worse. William asked Carr to capitalize on Dorothy's own wishes by a planned withdrawal of the doses. During September 1835, these were reduced by a half,[9] and by late November, gradual total withdrawal 'has nearly been effected without bad consequences'.[10] By mid-January 1836, Mr Carr gave it as his opinion that 'If abscesses do not form on the brain . . . she will live for years'. Carr's prognosis proved all too accurate: 'a melancholy prospect' her relatives now felt. 'If her mind is to continue in its present state,' wrote Dora to young Sara Coleridge, 'one and all of us would *joyfully* see her laid to rest . . . in our own quiet churchyard.'[11] Crabb Robinson, visiting in winter 1835, noted her 'offensive practises' and her sudden shouting or screaming 'without cause'. He wrote 'entre nous' to his brother that 'Poor Miss W. is sinking into imbecility',[12] and left Rydal Mount sadly, feeling that probably he would never see his old friend again. In fact she was to survive another twenty Christmases in a cruel caricature of old age.

Professional opinion suggests that Dorothy suffered presenile dementia of a type similar to Alzheimer's disease,* a degenerative disorder beginning in middle life, with a pathology distinct from the normal process of ageing. The underlying cause appears genetic, though this, in the Wordsworth or Cookson inheritance, has proved impossible to trace. The sharp-eyed De Quincey, however, made an observation which may be to the point. Both Dorothy and William showed such a 'premature old age . . . that strangers invariably supposed them fifteen to twenty years older than they were'. In hereditary cases, any small extra shock or damage to brain function may well contribute to dementia. Dorothy was already weakened by years of physical illness and opium. The arrival of Annette's letter, so full of a hurt long concealed, may have been the final shock which pushed her over the edge into confusion. In this group of diseases the processes of ageing are speeded and intensified; the brain atrophies and tangled filaments in the nervecells lose their function.[13] In the early stages patients lose memory for recent events, efficiency in everyday life, and orien-

* See Appendix Two.

-tation in time or space. The clean, meticulous person becomes disorderly and dirty, the self-effacing noisy and aggressive; Dorothy's irritation when her wishes were opposed amounted to 'rage and fury',[14] and William's patient attempts to soothe her left him shattered. As well as unpredictable changes of mood, her restless, purposeless activity, with demands or gestures repeated twenty times, wore out her family, though never their devotion. The mystery remains, of course, how Dorothy survived so long with a disease in which the expectation of life is commonly said to be only between five and ten years. Even allowing for possible remissions and the reserve capacity of her excellent brain, twenty years seems a long-drawn battle with this devastating disease.

One suggestion* is that she received, by the standards of any period, exceptionally good nursing care, with unfailing patience and affection from all around her. Her long years of self-sacrifice and devotion to the family were fully and generously repaid. William and Mary, with the radiant tenderness which had struck De Quincey so many years ago, gave loving care to their helpless dependant for the rest of her life. Mary Lamb had gone to a private mad-house and Mrs Southey to the Retreat at York, but there was no question of Dorothy leaving home; she remained in the comforting familiarity of Rydal Mount and its garden. Nor was she left lying, lonely and apathetic, in her room to die—like so many nineteenth-century invalids—from hypostatic pneumonia and bedsores. She was out of bed, dressed and moving, at first only dragging her feet between two attendants, but later reaching her bedroom window with a stick, 'so that we hope', wrote Mary, 'she will walk soon'. She had a bright fire in her room: ' "Stir the fire" is her first salutation . . . and that must either be done or a hubbub ensues.'[15] William came down from visiting her in January 1836 with tears in his eyes. 'Well,' he said, 'all I can do for her now is to heat her nightcap—I have done it 20 times within the last ¼ of an hour.' It was at least some comfort.[16]

When spring came, the garden gate was locked to shut out the curious, while the invalid chair with its wizened burden went round and round the green terraces. At the first sight of the spring flowers Dorothy wept, just as when a little girl she

---

* See Appendix Two.

had burst into tears at the sight of the sea. She asked to be taken to a favourite border but 'was at first too overcome to look at it,' reported Mary; 'on a sudden she began to sing'. She was out for two to three hours every morning and afternoon. Even while being dressed for the outdoors, she was heard singing away at *My Lady Cadogan*.[17] No song of this name appears in catalogues of printed, manuscript, or folk music; yet Mary wrote of it to Dora as something known to both. Possibly it was the work of the widowed Edward Quillinan, amateur poet and composer,[18] long familiar to Dorothy as a visitor to Rydal Mount. Certainly the song was a sign of pleasure.

Dorothy slept well as a result of the fresh air, and said she was 'never happy but when she was eating'. For the first time in years she gained weight, until she even grew very fat, seemed free from the familiar pains,[19] and eventually, for the first time since 1829 in Whitwick, 'as to bodily health perfectly well'. Cherished objects gave comfort and reassurance. Quillinan noted in his poems the rose-tree in the garden, planted in memory of Thomas, while outside her window grew a laburnum tree, where two doves hung for her delight in an osier cage. Above all, wrote Mary to Catherine Clarkson, 'it is a comfort to see that she is *happy*'.[20] Inevitably, to Mary's great distress, her personal habits deteriorated. Mary, characteristically, grieved not for herself, but for Dorothy's 'discomforts', of which she could not bring herself to speak. As with the outburst of profane or obscene speech, her only comment was a sad 'poor darling'. 'The continual washing will not be an easy matter in bad weather, but', she added matter-of-factly, 'it will be managed.'

Every evening Dorothy came downstairs to join the family circle. On good days, Crabb Robinson noted in his diary, 'she was fond of repeating the favourite small poems of her brother . . . in so sweet a tone as to be quite pathetic'.[21] On such days they could not believe they had lost her, and hoped against hope for a recovery. Attempts to interest her in reading, though, were a failure; she 'could not bear sustained attention to any book', explaining to Mary that she was 'too busy with her own feelings'. Letters to her were collected, unread—'She is too ill.' Nor did she care to hear reading aloud. On bad evenings she was 'more restless than ever', endlessly on the move

from her own chair to her brother's and back again with purposeless activity. 'And thus it is she wears away the day,'
reported Mary sadly. Yet, though others grieved for the loss of
her incomparable gifts, for Dorothy there may have been some
gain. In second childhood she found what was lost in her first,
the total security of family love for which she had hungered all
her life.

An unusual feature of Dorothy's illness was her continuing
ability to write mechanically, 'the act seems to require neither
time nor *thought*; she is so rapid'. Writing had been such an
essential part of her life that it apparently survived as an islet in
the general decay. On 4 November 1835, she had tried to 'take
up the pen again. . . . After a trying illness', but she was clearly
confused, and only managed a few disjointed sentences in the
journal, these being headed 'Nov. 4th, 1815', a slip-back of
twenty years.[22] On the other hand, sometimes her mind seemed
to clear itself in poetry. On 3 September 1836, she actually
wrote a short poem in the eighteenth-century diction and
couplets familiar from her youth. She describes relying on
others

> To prop my feeble steps, and try
> To soothe my pensive drooping heart
> With loving words, till we depart
> To seek within the parlour walls
> Comfort . . .

To Samuel Rogers, William had recorded that her mind 'is
much shattered', though when he read poetry to her she continued to quote it from memory,[23] a characteristic of many
senile patients. Mary observed, early in 1836, 'she *can* talk continually for a short time together, but otherwise her habits, etc,
are the same'.[24] These habits were violent, and she had to have
an attendant 'solely devoted to her', at this time a woman, also
called Dorothy, who was characterized by Mary early next
year (1837) as 'exemplary'. She needed to be; for in March
1836, in an unexpected letter of fierce wild humour to her
nephew Christopher, Dorothy wrote:

I have got through a mighty struggle—and thank God am now as well
as ever I was in my life except that I have not recovered the use of my
legs. My Arms have been active enough as the torn caps of my nurses
and the heavy blows I have given their heads and faces will testify.[25]

This bears out Mary's remark, 'She has much wild pleasure in her sallies.'[26] Dorothy ended the letter by announcing her intention, obviously impossible in her state, 'to see you all at Cambridge in summer'. Mary observed, when the summer came, that 'distressing as her state is—more especially to us who know what she once was—it is a comfort to see that she is *happy*; that is, she has no distress or sorrow that oppresses her more than the transient sorrow of a spoilt child'.[27] Inevitably, like any child, she had her griefs, when 'she is languid and weeps which is very afflicting,' wrote Mary to Robinson with unavailing sympathy.

Dorothy's moments of wild humour had already shown themselves to Crabb Robinson. Visiting for Christmas 1835 and New Year 1836 he was shocked to hear the once-pious Miss Wordsworth 'repeating the Doxology, and, the most melancholy act of all, carried away by the association of ideas, repeating the Amen in a loud tone and mocking a clerk accompanied by loud laughter'.[28] On the other hand, in October 1837, she astonished Mary and the household by writing a well-expressed and feeling letter to her Halifax cousin Edward Ferguson.[29] This was perhaps called out by the news that his and her 'Aunt', Mrs Rawson, was in her last illness. It began:

My dear Cousin Edward,

A Madman might as well attempt to relate the history of his doings and those of his fellows in confinement as I to tell you one hundredth part of what I have felt, suffered and done.

It ended:

I have not seen dear Charles Lamb's Book [of published Letters]. His Sister still survives—a solitary twig, patiently enduring the storm of life. In losing her Brother she lost her all—all but the remembrance of him—which cheers her the day through.

She added some of her own verses 'To Thomas Carr, my Medical Attendant', relating to her earlier physical crisis some years before in the winter of 1832-3. The letter shows that her remissions, when they came, were considerable. Yet, as Mary wrote about this time, 'alas these gleams are short lived'.

Another 'gleam' seems to have appeared a few months later in March 1838, perhaps in consequence of the death of Mrs Rawson, from whom she received a legacy. She wrote to Dora,[30]

who was staying away from Rydal Mount with Isabella Fenwick, the admirer and chronicler of William in his old age.

My dearest Dora,

They say I must write a letter—and what shall it be? News—news
I must seek for news. My own thoughts are a wilderness—'not
pierceable by power of any star'—News then is my resting-place
—news! news!
Poor Peggy Benson lies in Grasmere Church-yard beside her once
beautiful Mother. Fanny Haigh is gone to a better world. My Friend
Mrs. Rawson has ended her ninety and two years pilgrimage—and
I have fought and fretted and striven—and am here beside the fire.
The Doves behind me at the small window—the laburnum with its
naked seed-pods shivers before my window and the pine-trees rock
from their base.—More I cannot write so farewell! and may God
bless you and your kind good Friend Miss Fenwick to whom I send
love and all the best of wishes—Yours ever more

                                                    Dorothy Wordsworth

The ghost voice of her true self, speaking from the ruins, is
almost unbearably sad. It is typical that she should recollect
(though slightly misquote) a line from Spenser's *Faerie Queene*
for her mental state.

According to Mary, she spent much time 'pouring out verses'.
These were 'generally addressed to her attendants—the subjects are not very elevated'. She also read old newspapers, and
seemed 'generally happy and good-tempered', though she 'with
the impatience of a *petted child* contrives one *want* after another'.[31]
Her head was shaven, according to the contemporary idea that
the heads of the insane should be kept cool, and Mary recorded,
in 1838, her happier moments, 'generally with a basin of hot
water, in which she washes her hands and shaven head—upon
her knee—amusing herself with her nail brush—and soap lather
reaching to her chin'. Mary commented with relief that Dorothy
'seldom uses bad expressions except in fun', though, when the
house was full of visitors, 'We can give *her* no neighbours but
ourselves, or she would terrify strangers to death.' Sadly, children, whom she had so greatly loved, were frightened of her,
but she did not 'much care'. In 1840, she was notably amused
by the sound and sight of the family cuckoo-clock. 'I thought
she would have dropped from her chair, she laughed so heartily
at the exit of the little Mimic', but other amusements were 'not

very elevated'. Robinson noted in 1841:

She has so little command of herself that she cannot restrain the most
unseemly noises, blowing loudly, & making a nondescript sound
more shrill than the cry of a partridge or a turkey.[32]

A visitor to a modern geriatric ward may often hear sounds
such as these, and find them as distressing as Robinson clearly
did.

One wonders if she was really conscious of a change which
took place around her at Rydal Mount in that same year, 1841,
when Dora married the widower Quillinan, whose first wife
Dorothy, years ago, had nursed. The agitation of this event to
William was considerable, and he was only reconciled to the
marriage by the good offices of Miss Fenwick. Quillinan now
became one of the family, and his comments on Dorothy are
made from close observation. In the next year, 1842, he gave it
as his opinion of her that 'she is very well and oftener merry
than sad'. The subjects of her merriment are not stated, nor
are her reactions, if any, when in 1843 William succeeded
Southey as Poet Laureate. Quillinan comes strangely into that
picture, since he wrote the new Laureate's Ode to the Prince
Consort in 1847 at a time when Dora, after only six years' marriage, lay dying. Her father was so broken with grief that he
could not compose, and so her husband had to help him. The
only comfort William seemed to find, Robinson noticed, was
caring for his sister. 'Her death to him would be a sad calamity.'
Mary in silent grief accepted Dora's death and Dorothy's survival without question, as the mysterious will of God. Of all
this irony, Dorothy seemed unaware. A local artist, John
Harden of Brathay, who sketched her, sharp-nosed and swad-
dled in her wheelchair, captured the isolation of the alienated.

From time to time, other visitors had brief, chance meetings.
One of these, calling at the Rydal Mount front-door in 1849,
was Julia Wedgwood, whose great-uncle Tom Wedgwood had
stayed for some days under Dorothy's roof at Alfoxton over fifty
years before. She wrote:

I was waiting at the door when her chaise drew up, bearing the little
shrunken figure from her daily excursion, and I looked into those
'wild eyes' which kept all their life and light though the mind had
grown dim. There was no dimness in her interest when she heard my
name. 'From whom are you sprung?' she enquired eagerly.[33]

When, in the following year, William himself lay dying, Quillinan wrote to Robinson of the unexpected effect on her of her Brother's impending death.

Miss W. is as much herself as she ever was in her life, & has an almost absolute command of her own will! does not make noises; is not all self; thinks of the feelings of others (Mrs W's for example), is tenderly anxious for her brother; and in short but for age and bodily infirmity, is almost the Miss Wordsworth we knew in past days.[34]

What seemed so surprising to Quillinan was a fact observed by students of senile dementia, that a strong stimulus may suddenly make the patient much more accessible. Nor did it last. Only a few months after William died, Harriet Martineau commented, 'If that poor sister were released, it seems as if it would be a great blessing.'

There were other occasions when she was 'not all self'. 'My brothers were all good men, good, good!' she was heard to say,[35] adding almost in the words of her first letter to Jane, long years ago, 'The boys in our family are all good.' In the years that remained to her, there were evidently occasional respites. Though Mary could write of her, early in 1853, that she was 'in her usual state—yet more indifferent to gone by Events. She passes more of her time in bed: which is a relief to her Attendant as well as herself', there were unusually lucid moments. Later the same year, when Mary herself was seeking relief in a visit to Miss Fenwick, Dorothy after eighteen years' illness could write, coherently if a little childishly:

My dearest Sister,

I have had a good night so I think I will write. The weather was rough. I was in bed all day. I am well today. My love to Miss Fenwick and Miss Jane—love to Hanna.

Dorothy Wordsworth

Mrs. Pearson is very poorly and the Doctors say she cannot live. We have got a cow and very good milk she gives. I only wish you were here to have some of it.

Thomas Flick is a little better and we are all quite fit. Mary Fisher's Sister is just dead.[36]

These were her last recorded words. In July 1854, Mary wrote to Crabb Robinson, 'Your old friend is very much disturbed—so much as it has been very sad to see her.' The end came on 25

January 1855, peacefully. There were present with her Mary, her two nephews, John and Willy, John's two boys, Mr Carr, and James Dixon the family servant.

Mary wrote next day to break the news to the Hutchinsons, with whom Dorothy had passed such idyllic months at Brinsop Court. 'Our dear Sister was released after her gradual but *fitful* sinking and some few hours of peaceful and anxious waiting.' They need not feel anxious on her own account. 'You are with me in *heart* and that is all I wish for.' On 7 February 1855, a bright morning, she wrote to Susan, the wife of Dorothy's nephew Christopher Wordsworth. There would always be a 'void' at her heart, she admitted, yet how much better for her to survive alone. 'I shall forever feel thankful for the Almighty's goodness for having spared me to be the solitary lingerer, rather than the beloved sufferer now laid at rest and whose restless Spirit I humbly trust is now among the blessed in the bosom of her heavenly Father.'[37]

A whole generation had grown up knowing Dorothy only as 'poor Miss Wordsworth'; yet 'even so,' wrote one of them— Matthew Arnold's young brother—'the thought of her lying beside her Brother in Grasmere church yard is touching'. Mary lived four more years, sight fading but clear of spirit, before joining them 'perfectly happy'. In the quiet churchyard the graves lie within the river's arm. Dorothy is surrounded in death by the lives which had shaped her own: the two children so bitterly mourned in 1812, the old friend and companion Sara gone in 1835, Dora, stubborn fighter for life, come home to die in 1847, William buried with poet's honours in 1850, and Mary, last faithful survivor of the band, in 1859. Her best memorial, her journals, lived on in family possession, full of phrases which might well describe her own life and death: not least among them, her epitaph on the pauper woman she had seen buried in the same churchyard.

I thought she was going to a quiet spot.

Appendix One

Letter from Annette Vallon to
Dorothy Wordsworth, April 1835

Mademoiselle Dorothée Wordsworth
Grasner Ridelmonde
Kendal
Westmorland
angleterre
                              [Date stamp of arrival Ap 4 1835]

C'est avec plaisir et chagrin que je vous ecris ma bien chere Dorothée
ses sentimens opposés n'en sont pas moins ces que jeprouve j'attendois
une reponse á la lettre que jai ecrit á votre frere j'esperois une pronte
satisfaction a mes justes reclamations il prolonge mes souffrances, et
le chagrin qui me tue, cette pensée doit le tourmenté parce je le
suppose toujours avec cette noblesse de coeur que je lui ai naguere con-
nu et constament admiré. Sa position de famille peu l'embarassé mais
la mienne doit le toucher et sa fille doit le déterminé á acquitté une
promesse faite; ce qu'il envoye ne represente point á son mari la
somme donnée par son contra de marriage; je ne rescenderai sur des
détails deja donné dans ma precedent lettre. Le passé avec tous les
souvenirs qu'il me laisse n'a rien que d'affligent, le present m'occupe
par amour pour ma fille et son avenir me tourmente Si vous ne deter-
miné pas votre frere á remplir noblement ce que je crois qu'il doit
tant de motifs d'honneur et délicatesse lui en sont la loi que je
presume que vous n'aurai pas de peine a le décidée a faire cesser cet
etat painible, les fonds sont sans interet ils reposent entre les mains
du Banquer sans nulle benifite pour M. Beaudoin il est bon mari bon
pere mais cest en raison de ces qualités qu'il est plus strique á
demander pour eux.

je vous le demande au nom de l'amitié que je vous ai voue et au
nom de mes malheurs qui doivent parlé haut dans son coeur de ter-
miner ces reclamations si painible pour moi. Ma chere fille si admirable dans les qualités que possede son père merite toute sa tendresse
par la ressemblance morale qu'elle a avec lui, elle me charge de lui
offrire toute l'etendue de son respect de sa tendresse elle lembrasse
avec toute la chaleur de son coeur et vous Bonne Dorothée qu'elle
aime tant en qui elle á la plus grande confiance justifié ce sentiment
en prenant les interets de votre niece et laissé mourir sa mere en paix
je ne tiens guere á la vie dieu peut en abregé la cour quant il lui
plaira.

Vos nieces sont charmantes elles sont parfaitment élevées la jeune vas faire sa première communion ma derniere invoquation sadresse bien vivement a votre frere Cest entre ces mains que reposent les derniers jours de ma carriere.

Je vous embrasse avec de sentiments qui ne peuvent changer.

Mes respects á Madame Words qu'elle sois mon avocat je serai sure du sucès

adieu votre Malheureuse Annette W

2 avril 1835   Boulevard des Filles du Calvaire au 15

[Annette's spelling is not grammatical but phonetic: for example, 'é' does service for both '-er' and '-ez'.]

Appendix Two

Dorothy Wordsworth's Medical Condition

(Reproduced by kind permission of Dr I. I. J. M. Gibson, FRCP, Consultant Physician in Geriatric Medicine, Southern General Hospital, Glasgow.)

In reviewing Dorothy Wordsworth's mental state, the appreciation that this could be a toxic condition, resulting from gallstones and gallbladder inflammation, has been considered. However, I think we try too much to link all her illnesses, and in fact that Dorothy Wordsworth had senile dementia of the type similar to Alzheimer's disease which is a genetically determined pre-senile dementia. This is a disorder coming on usually between 40 and 60 but which in no other way differs from senile dementia. The pathological picture in the brain is the same, but for some reason striking at a much earlier age.

Mary Wordsworth really gives the most clear of clinical pictures. She describes a person who has had excellent memory now dwindled away, a person who evades questions that she cannot answer by saying that she is 'involved in her own feelings' which is a typical evasion of dementia. She describes a patient who can walk but does not and yet believes that she is actually walking; who is facile, emotional, childish, given to rages, but who can be soothed and will laugh at the cuckoo clock or play quite happily with the basin of water. The Wordsworths found it a great problem that she could recite William's verses, given a start, and go on without error. Repetition of childhood or well-loved verses is quite frequent among the demented, sometimes seen as the ability to recall and sing hymns, often with many verses. Dorothy was of course conditioned to William's poetry and it is no surprise that this islet remained. Restlessness, wandering, deterioration of personal habits and cleanliness are all obvious things in a diagnosis of dementia.

The flashes of personality that come through are again not incongruous particularly if the person has been quick witted before and has special interests. I have however always been troubled by the letters she wrote during her demented period, since this is unusual. Note Mary's comment, 'the accompanying letter from Dorothy cannot surprise you more than it has done us—since her confinement she has never before fixed her mind for so long I think to one effort'. I think this flash of intellect is just a more prolonged revelation of the basic personality than the odd isolated remarks that she would make.

When I first started work in Geriatric Medicine I learnt that senile dementia of moderate severity would lead to death within probably two to three years. I think we now realize that dementia when the patient is adequately cared for has a much longer time span. Dorothy did survive an unusually long time. Mary often says that she is relatively happy and she did receive care and constant attention. There is no doubt at all that she was maintained in life by the loving care given to her by her family and by the possibility of employing servants to ensure her physical care. The Wordsworths were really practising good modern psycho-geriatric medicine. Compare and contrast the care of King George III.

Notes

ABBREVIATIONS USED IN NOTES

CN	*The Notebooks of Samuel Taylor Coleridge*, 3 double vols. ed. Kathleen Coburn, 1957–.
DC	Dove Cottage MSS in the Wordsworth Library, Grasmere.
De Selincourt	E. de Selincourt, *Dorothy Wordsworth*, 1933.
DWJ	*The Journals of Dorothy Wordsworth*, ed. Ernest de Selincourt, 2 vols., 1941.
HCR	*Henry Crabb Robinson on Books and their Writers*, 3 vols., ed. E. J. Morley, 1938.
KC	*The Keats Circle*, ed. H. E. Rollins, 2 vols., 1948.
Legouis	E. Legouis, *William Wordsworth and Annette Vallon*, 1922.
Letters	*The Letters of William and Dorothy Wordsworth*, second ed., revised, 6 vols., 1967–.
Letters C and ML	*The Letters of Charles and Mary Lamb*, 3 vols., ed. E. V. Lucas, 1935.
Letters W and MW	*The Love Letters of William and Mary Wordsworth*, ed. Beth Darlington, 1982.
LMW	*Letters of Mary Wordsworth*, ed. M. E. Burton, 1958.
LSH	*Letters of Sara Hutchinson*, ed. Kathleen Coburn, 1954.
Lindop	G. Lindop, *The Opium Eater: a Life of Thomas De Quincey*, 1982.
Moorman	M. Moorman, *William Wordsworth, A Biography*, I, The Early Years, II, The Later Years, 1957 and 1965.
PW	*The Poetical Works of William Wordsworth*, 5 vols., ed. E. de Selincourt and H. Darbishire, 1940–.
Reed	*Wordsworth: the Chronology of the Early Years*, ed. Mark L. Reed, 1967.
Southey, *Letters*	*The Letters of Robert Southey, a Selection*, ed. M. H. Fitzgerald, 1912.
STC Letters	*Collected Letters of Samuel Taylor Coleridge*, 6 vols., ed. E. L. Griggs, 1956–71.
T. De Quincey, *Recollections*	*Recollections of the Lakes and the Lake Poets*, by Thomas De Quincey, ed. D. Wright, 1970.
W Circ.	*Correspondence of Henry Crabb Robinson with the Wordsworth Circle*, 2 vols., ed. E. J. Morley, 1927.
WL	The Wordsworth Library, Grasmere.

CHAPTER 1

1. *Letters*, i, No. 277.
2. *The Prelude* (1850), i, 275–94.
3. *DWJ* i. 413.
4. *Letters*, ii, No. 83.
5. *The Prelude* (1805–6), i. 257–8.
6. *Letters*, i, No. 257.
7. *Letters*, i, No. 256.
8. Moorman, i. 14.
9. *The Prelude* (1805-6), v. 262.
10. C. Maclean, *Dorothy Wordsworth: The Early Years*, p. 8, gives no reference, but 'some Extracts from the Diary of a Halifax Lady' were reprinted in the *Halifax Guardian*, 1887–9. The diarist, Anne Lister of Shibden Hall, knew Dorothy Wordsworth's cousin well in her old age and several points in *Dorothy Wordsworth: The Early Years* are confirmed by her entries.
11. *Halifax Guardian*, 6 July 1889, p. 6, c. 4.
12. *Letters*, i. No. 1, p. 2, n. 2.
13. *Letters*, i, No. 30.
14. *Letters*, i, No. 12.
15. *Letters*, i, No. 65.
16. MS Journals, Wednesday, 25 June [1834], DC 118, WL.
17. P. Darby, *An Essay on Halifax* (1761).
18. F. E. Millison, *Two Hundred Years of the Northgate End Chapel* (1896), and R. Eccles, *Northgate End Chapel, Halifax* (1946).
19. *Letters*, i, No. 11.
20. *Letters*, i, No. 280.
21. *Letters*, i, No. 9.
22. *Letters*, i, No. 31.
23. *Letters*, iii, No. 359.
24. *Letters*, i, No. 293.
25. *Letters*, ii, No. 92.
26. *Letters*, i, No. 11, p. 42, n. 3.
27. *Letters*, i, No. 35.
28. *Letters*, i, No. 14.
29. *Letters*, i, No. 12.
30. *Letters*, i, No. 4.
31. D. M. Stenton, *The English Woman in History*, p. 278.
32. *Letters*, i, No. 8.
33. *Letters*, ii, No. 177.
34. *Halifax Guardian*, 13 April 1889, p. 6, c. 5.
35. *Letters*, ii, No. 80.
36. *Letters*, v, No. 555.
37. *Halifax Guardian*, 27 August 1887, p. 6, c. 4.
38. *Letters*, ii, No. 81.
39. *Letters*, i, No. 2.
40. *Letters*, i, No. 50.
41. *Letters*, i, No. 19.
42. *Letters*, i, No. 9.

CHAPTER 2

1. *Letters*, i, No. 5.
2. *Letters*, i, No. 3, p 10, n. 2, suggests that Dorothy Wordsworth attended Penrith Grammar School. She knew the headmaster, but grammar schools, which taught Latin and Greek grammar to boys, did not admit girls as students.
3. *Letters*, i, No. 3.
4. *Letters*, i, Nos. 19 and 49.
5. *Letters*, i, No. 5.
6. *Letters*, i, No. 1.
7. *Letters*, i, No. 1.
8. *Letters*, ii, No. 67.
9. *Letters*, No. 3.
10. *Letters*, i, No. 293.
11. *Letters*, i, No. 14.
12. *Letters*, i, No. 1.
13. *Letters*, i, No. 50.
14. *Oxford History of English Literature*, VIII, 465.
15. *Letters*, i, No. 5.
16. *Letters*, i, No. 30
17. *Letters*, i, No. 3.
18. *Letters*, i, Nos. 33 and 34.
19. *DWJ* i. 421.
20. William Hutchinson, *History and Antiquities of Cumberland*, 313–44.
21. *DWJ* i. 420.
22. *Letters*, i, No. 2.
23. *Letters*, i, No. 4.
24. *Letters*, i, No. 2.
25. *Alumni Cantabrigiensis*.
26. *Letters*, i, No. 3.
27. *Letters*, i, No. 4.
28. *The Prelude* (1805), vi. 210–36.
29. *Letters*, i, No. 6.
30. *Letters*, i, No. 6.
31. *The Prelude* (1805), iii. 36–43.

CHAPTER 3

1. Blomefield, *Topographical History of Norfolk*, vol. v (1806), p. 223.
2. James Woodforde, *Diary*, 4 Jan. 1789.
3. *Letters*, i, No. 6.
4. *Letters*, i, No. 7.
5. *Letters*, i, No. 8
6. *Letters*, i, No. 19
7. *Letters*, i, No. 9.
8. *Letters*, i, No. 8.
9. *Letters*, i, No. 8.

10. Sir F. Eden, *The State of the Poor* (1795), Parochial reports: Gressenhall, Norfolk.
11. *Letters*, i, No. 288.
12. *Letters*, i, No. 9.
13. *Letters*, i, No. 12.
14. *Letters*, i, No. 7.
15. *Letters*, i, No. 19.
16. *Letters*, i, No. 14.
17. *Letters*, i, No. 11.
18. *Letters*, i, No. 11.
19. *Letters*, i, Nos. 9 and 11.
20. *Letters*, i, No. 171.
21. *Letters*, i, No. 50.
22. *Letters*, i, No. 19.
23. *Letters*, i, No. 22.
24. *Letters*, i, No. 24.
25. *Letters*, i, No. 30.
26. *Letters*, i, No. 31.
27. *Letters*, i, No. 14.
28. M. Pennington, *Memoirs of the Life of Mrs. Elizabeth Carter* (1807).
29. *Letters*, i, No. 50.
30. *Letters*, i, No. 26.
31. *DWJ* ii. 86.
32. *Letters*, i, No. 8.
33. *Letters*, i, Nos. 5 and 8.
34. *Letters*, i, No. 28.
35. *Letters*, i, No. 26.
36. *Letters*, i, No. 14.
37. *Letters*, i, No. 21.
38. *Letters*, i, No. 12.
39. William Hazlitt, quoted Moorman, i. 398.
40. *Letters*, i, No. 14.
41. *Letters*, i, No. 12.
42. *Letters*, i, No. 28.
43. *Letters*, i, No. 30.

CHAPTER 4

1. *Letters*, i, No. 16.
2. *Letters*, i, Nos. 17 and 18.
3. *Letters*, i, No. 16.
4. *Letters*, i, No. 23.
5. *Letters*, i, No. 4.
6. *Letters*, i, No. 24.
7. *Letters*, i, No. 28.
8. *Letters*, i, No. 28.
9. *Letters*, i, No. 22.
10. *Letters*, i, No. 20.

11. Legouis, p. 18.
12. *Letters*, i, No. 25.
13. *Letters*, i, No. 27.
14. *Letters*, i, No. 31.
15. Legouis, p. 33 and Appendix.
16. *Letters*, i, No. 29.
17. *Letters*, i, No. 30.
18. *Letters*, i, No. 31.
19. *Letters*, i, No. 35.
20. *Letters*, i, Nos. 33 and 34.
21. *Letters*, i, No. 30.
22. *Letters*, i, No. 31.
23. *Letters*, i, No. 35.
24. *Letters*, i, No. 36.
25. *Letters*, i, No. 50.
26. *Letters*, i, No. 31.
27. *Letters*, i, No. 33.
28. James Woodforde, *Diary*, 23 June 1786: 'For 3 Peoples Fare to London I pd. 4.10.0', and Lowndes, *Guide to Stage Coaches* (1794).
29. *Letters*, i, No. 36.
30. *Letters*, i, No. 31.
31. *Letters*, i, No. 40, n. 4.
32. *Letters*, i, No. 41.
33. *Letters*, i, No. 40.
34. *Letters*, i, No. 42.
35. *Gentleman's Magazine*, May 1794.
36. *DWJ*. i. 182.
37. *DWJ* i. 104.
38. *Letters*, i, No. 38.
39. *Letters*, i, No. 37.
40. *Letters*, i, No. 39. This letter was discovered in a box lying in an outhouse at the Crackenthorpes' house, Newbiggin Hall.
41. *Letters*, i, No. 277.
42. *Letters*, i, No. 41.
43. *STC Letters*, i, No. 351.
44. *Letters*, i, No. 38.
45. *Letters*, i, No. 49.
46. *Letters*, i, Nos. 48 and 49, n. 1.
47. *Letters*, i, No. 49.
48. *Letters*, i, No. 49.
49. *Letters*, i, No. 50.
50. *Letters* iv, frontispiece.
51. C. Maclean, *Dorothy Wordsworth: The Early Years*, facing p. 259.

CHAPTER 5

1. *Letters*, i, No. 50.
2. *Letters*, i, Nos. 44 and 45.

3. *Letters*, i, No. 44, n. 3.
4. Moorman, i. 269–70.
5. Ibid.
6. *Letters*, i, No. 50.
7. De Selincourt, pp. 58–9.
8. *Letters*, i, No. 50.
9. *Letters W and MW* p. 61.
10. Reed, p. 169, n. 11.
11. Sir F. Eden, *The State of the Poor* (1795), Book ii, ch. 11.
12. *Letters*, i, No. 55.
13. *Letters*, i, No. 127.
14. *Letters*, i, No. 56.
15. *Letters*, i, No. 58.
16. *Letters*, i, No. 55.
17. *Letters*, i, No. 58.
18. *Letters*, i, No. 65.
19. *Letters*, i, No. 93.
20. *Letters*, i, No. 58.
21. *Letters*, i, No. 55.
22. *Letters*, i, No. 55.
23. *Letters*, i, No. 127.
24. *Letters*, i, No. 58.
25. Moorman, i. 267, n. 3.
26. Moorman, i. 280.
27. *Letters*, i, No. 55.
28. Sir F. Eden, *The State of the Poor* (1795), Parochial reports; Clyst St. George.
29. Moorman, i. 320, n. 2, though the story is doubtful.
30. De Selincourt, p. 64.
31. *Letters*, i, No. 58.
32. *Letters*, i, No. 65.
33. *Letters*, i, No. 75.
34. *Letters*, i, No. 62.
35. *DWJ*, i. 243–4.
36. Legouis, p. 56.
37. *Letters*, i, No. 140.
38. *Letters*, i, No. 50.
39. *Letters*, i, No. 67.
40. *Letters*, i, Nos. 66 and 69.

CHAPTER 6

1. *Letters*, i, No. 70.
2. *Letters*, i, No. 70.
3. *STC Letters*, No. 195.
4. *STC Letters*, No. 197.
5. J. D. Campbell, *Samuel Taylor Coleridge*, p. 75.
6. *STC Letters*, No. 422.

7. *STC Letters*, No. 393.

8. *CN* i. 570.

9. William Hazlitt, 'My First Acquaintance with Poets'.

10. *STC Letters*, i, No. 192.

11. *Letters*, i, No. 71.

12. *DWJ* i. 251.

13. *Letters*, i, No. 72.

14. Ibid.

15. *STC Letters*, No. 197.

16. 'A Lady' [Hannah Glasse], *The Art of Cookery made Plain and Easy*, 21st edn. (1796).

17. *DWJ* i. 403.

18. *Letters*, i, No. 72.

19. Joseph Cottle, *Reminiscences of Coleridge and Southey*, p. 181.

20. A. J. Eagleston, 'Wordsworth, Coleridge and the Spy', in *Coleridge*, ed. Edmund Blunden and E. L. Griggs. pp. 73–87.

21. Coleridge, *Biographia Literaria*, ch. X.

22. Mrs H. Sandford, *Thomas Poole and his Friends*, i. 242.

23. Ibid.

24. William Hazlitt, 'On Going a Journey'.

25. *CN* i. 286.

26. *Letters*, i, No. 72.

27. *STC Letters*, No. 1621 and n. 4.

28. *STC Letters*, No. 209. E. L. Griggs's dating seems by far the most likely.

29. *DWJ* i. 34.

30. *Letters*, i, No. 76.

31. Cf. William Hazlitt, 'My First Acquaintance with Poets'.

32. *Letters*, i, No. 76.

33. Coleridge, 'The Wanderings of Cain'. The resemblance between Coleridge's synopsis and certain features of Wordsworth's poem is striking.

34. *Letters*, i, No. 75. William did not pay for another fifteen years.

35. *Letters*, i, No. 77.

36. *PW* i. 361.

37. *Ancient Mariner* quotations as in *Lyrical Ballads* (1798).

38. *Letters*, i, No. 49.

39. *STC Letters*, No. 212.

40. *Letters*, i, Nos. 79, 80 and n.

41. *DWJ* i. 278.

CHAPTER 7

1. See below, p. 100.

2. De Selincourt, p. 78.

3. *Letters of John Keats*, ed. H. E. Rollins, ii. 89.

4. *CN* i, 318.

5. *CN* iii. 4007.

6. *Letters*, i, No. 90.

7. *Letters*, i, No. 84.

8. *Letters*, i, No. 83.
9. Moorman, i. 370–1.
10. *Letters*, i, No. 85.
11. *Letters*, i, No. 88.
12. *STC Letters*, i, No. 248.
13. *Letters*, i, Nos. 87 and 92.
14. *DWJ* i. 16.
15. William Hazlitt, 'My First Acquaintance with Poets' and 'On Going a Journey'.
16. J. Cottle, *Reminiscences of Coleridge and Southey*, pp. 182–5.
17. *Letters*, i, No. 91.
18. *Letters*, i, No. 77.
19. *Letters*, i, No. 120.
20. *Letters*, i, No. 93.
21. *Letters*, i, No. 94.
22. *Letters*, i, No. 95, p. 226, n. 1.
23. *Letters*, i, No. 93, p. 222, n. 1.
24. *Letters*, i, No. 96.
25. *Letters*, i, No. 97.
26. *CN* i, 335.
27. *DWJ* i. 20.
28. *DWJ* i. 24.
29. *CN* i. 346.
30. *Letters C and ML* i. 141.
31. *DWJ* i. 29-31.
32. *Letters*, i, No. 102.
33. *Letters*, i, No. 109.
34. *Letters*, i, No. 102, n. 1.
35. *Letters*, i, No. 103.
36. *STC Letters*, No. 270.
37. *Letters*, i, No. 106.
38. *STC Letters*, No. 266.
39. Moorman, i. 425.
40. *Letters*, i, No. 118.
41. Moorman, i. 430–1.
42. *Letters*, i, No. 110.
43. *STC Letters*, No. 276.
44. *STC Letters*, Nos. 276 and 277.
45. *Letters*, i, No. 105.
46. *Letters*, i, No. 123.
47. *Letters*, i, No. 140.
48. *Letters*, i, No. 113.
49. Alan G. Hill, 'Wordsworth and the Two Faces of Machiavelli', *Review of English Studies*, xxxi, No. 123 (1980).
50. *STC Letters*, No. 289.
51. *Letters*, i, No. 158.
52. *STC Letters*, No. 299.
53. *Letters*, i, No. 124.

CHAPTER 8

1. T. De Quincey, *Recollections*, p. 122.
2. *Letters*, i, No. 126.
3. *DWJ* i. 39, n. 3.
4. *Letters*, iii, No. 378.
5. *Letters*, i, No. 140.
6. *Letters*, i, No. 213.
7. *Letters*, i, No. 126.
8. *Letters*, i, No. 288.
9. *Letters*, i, No. 255.
10. *Letters*, i, No. 129.
11. See above, p. 76.
12. *STC Letters*, No. 328.
13. *STC Letters*, No. 340.
14. *Letters*, i, No. 140.
15. See above, p. 77.
16. *CN* i. 1782.
17. *DWJ* i. 90.
18. DC 118 WL.
19. *Letters*, i. No. 257.
20. *DWJ* i. 142.
21. *DWJ* i. 62–3.
22. *DWJ* i. 39.
23. *DWJ* i. 44.
24. *Letters*, i, No. 140.
25. *STC Letters*, No. 328.
26. *DWJ* i. 53.
27. *DWJ* i. 62.
28. *Letters*, i, No. 203.
29. *DWJ* i. 59.
30. *DWJ* i. 69.
31. Houghton Keats Room 4.20.11 79.
32. *Letters*, i, No. 131.
33. *Letters*, i, No. 272.
34. F. P. Rand, *Wordsworth's Mariner Brother*, p. 61.
35. *Letters*, i. No. 140.
36. *DWJ* i. 100.

CHAPTER 9

1. W. Knight, *Journals of Dorothy Wordsworth*, Prefatory Note.
2. *Letters*, i, No. 160.
3. *DWJ* i. 113.
4. *Letters*, ii, No. 80.
5. *Letters*, i, No. 161.
6. *DWJ* i. 48.
7. *DWJ* i. 46.

8. *Letters*, i, No. 140.
9. *DWJ* i. 54.
10. *Letters*, i, No. 182.
11. *DWJ* i. 83.
12. *DWJ* i. 85.
13. *Letters W and MW*, p. 203.
14. *DWJ* i. 188.
15. *Letters*, i, No. 140.
16. *DWJ* i. 120.
17. *DWJ* i. 49.
18. *DWJ* i. 54.
19. *DWJ* i. 60.
20. *Letters*, i, No. 140.
21. *Letters*, i, No. 140.
22. *DWJ* i. 79.
23. *DWJ* i. 64.
24. *DWJ* i. 233.
25. *DWJ* i. 38.
26. *DWJ* i. 43.
27. *DWJ* i. 42.
28. *DWJ* i. 50.
29. *Letters*, i, No. 186.
30. *DWJ* i. 46.
31. Keith Thomas, *Man and the Natural World*, p. 234.
32. Sir F. Eden, *The State of the Poor*, (1795), pp. 99, 127.
33. *DWJ* i. 40.
34. *DWJ* i. 38.
35. *DWJ* i. 47.
36. *DWJ* i. 85.
37. *DWJ* i. 86.
38. *DWJ* i. 93.
39. *DWJ* i. 120.
40. *DWJ* i. 124.
41. *DWJ* i. 112.
42. *Letters*, ii, No. 148.
43. *Letters*, ii, No. 223.
44. *DWJ* i. 43.
45. *DWJ* i. 133.
46. *DWJ* i. 59.
47. *DWJ* i. 234.
48. *DWJ* i. 112.
49. *DWJ* i. 140.
50. *DWJ* i. 126-7.
51. *DWJ* i. 181.
52. *DWJ* i. 147.
53. *Letters C and ML* i. 241.
54. *Letters*, i, No. 160.
55. *Letters*, i, No. 204.

56. *Letters*, i, No. 160.
57. *Letters*, i, No. 163.
58. *Letters*, i, No. 161.
59. *Letters*, i, No. 169.

CHAPTER 10

1. *Letters*, i, No. 164.
2. I. I. J. M. Gibson, *British Medical Journal*, vol. 285, 18–25 Dec. 1982.
3. *Letters*, i, No. 205.
4. *Letters*, i, No. 282.
5. *DWJ* i. 79.
6. Ibid.
7. *DWJ* i. 84.
8. *DWJ* i. 80.
9. *DWJ* i. 94.
10. *DWJ* i. 92.
11. *DWJ* i. 98.
12. *DWJ* i. 99.
13. *DWJ* i. 102.
14. *DWJ* i. 116.
15. *Letters W and MW*, p. 212.
16. *DWJ* i. 128.
17. *CN* iii, No. 3304.
18. *DWJ* i. 130.
19. *DWJ* i. 131.
20. *Letters*, i, No. 169.
21. *DWJ* i. 132.
22. *DWJ* i. 136.
23. *STC Letters*, No. 449.
24. See above, p. 57.
25. *DWJ* i. 128
26. *DWJ* i. 137.
27. *DWJ* 142–3.
28. *Letters*, i, No. 172.
29. Moorman, i. 543.
30. *Letters*, i, No. 172.
31. *DWJ* i. 78.
32. *Letters*, i, No. 171.
33. *DWJ* i. 157, and *Letters*, i, No. 172.
34. *DWJ* i. 159.
35. *Letters*, i, No. 171.
36. *DWJ* i. 168.
37. *DWJ* i. 169–71.
38. *LSH*, No. 2.
39. *Letters*, i, No. 171.
40. *DWJ* i. 172–3.
41. *DWJ* i. 174.

42. Legouis, p. 54; see Legouis, Appendix IV, for petition of 1816, signed by 20 referees, on Annette's services to Royalism.
43. Legouis, p. 99.
44. *DWJ* i. 174–5.
45. *DWJ* i. 175.
46. *Letters C and ML* i. 312.
47. *Letters C and ML* i. 246.
48. *Letters*, i. No. 175.
49. *Letters*, i. No. 177.
50. *Letters*, i, No. 178.
51. *DWJ*, ed. Moorman, p. 154 n. This passage was erased in the manuscript, though it may merely have recorded a thoughtful precaution by Dorothy. A man said to have mislaid a horse was not the ideal guardian for a ring.
52. *DWJ* i. 176.
53. *Letters*, i, No. 176 and p. 375, n. 2.
54. Humphry House, *All in due Time*, p. 27.
55. *DWJ* i. 176.
56. *CN* 1250.

CHAPTER 11

1. *DWJ* i. 183.
2. T. De Quincey, *Recollections*, p. 188.
3. *Letters W and MW*, p. 151.
4. *Letters*, ii, No. 29.
5. *Letters W and MW*, p. 167.
6. *Letters W and MW*, pp. 82 and 229.
7. *Letters W and MW*, p. 157.
8. *Letters*, i, No. 140.
9. *Letters*, i, No. 180.
10. *Letters*, i, No. 180.
11. F. P. Rand, *Wordsworth's Mariner Brother*, p. 26.
12. *Letters*, i, No. 183.
13. *Letters*, i, No. 184.
14. *DWJ* i. 188.
15. *Letters*, i, No. 197.
16. *Letters*, i, No. 196.
17. *Letters C and ML* i. 352–4.
18. T. De Quincey, *Recollections*, p. 205–6.
19. *Letters*, i, No. 221.
20. *Letters*, i, No. 189.
21. *Letters*, i, No. 192.
22. *Letters*, i, No. 193.
23. *DWJ* i. 259.
24. *Letters*, i, No. 192.
25. Ibid.
26. *DWJ* i. vii.

27. *DWJ* i. ix.
28. *STC Letters*, No. 574.
29. *DWJ* i. 277-8.
30. A. Dyce, *Recollections of the Table-Talk of Samuel Rogers*, pp. 208–9.
31. *DWJ* i. vii.
32. See above, p. 72.
33. *Letters*, i, No. 192.
34. *Letters*, i, No. 217.
35. *Letters*, i, No. 277.
36. *Letters*, i, No. 287.
37. *DWJ* i. 289-90.
38. *CN* i. 1463.
39. *STC Letters*, Nos. 321, 324, 525.
40. *Letters*, i, No. 200.
41. *STC Letters*, No. 539.
42. *CN* i. 1755.
43. *CN* i. 1830.
44. *Letters*, i, No. 223.
45. *Letters*, i, No. 226.
46. *Letters*, i, No. 229.
47. *Letters*, i, No. 226.
48. *Letters*, i, No. 231.
49. *Letters*, i, Nos. 219 and 222.
50. *Letters*, i, No. 199.
51. *Letters*, i, Nos. 244 and 245.
52. *Letters*, i, No. 288.
53. WL MSS.
54. *Letters*, i, No. 293.
55. *Letters*, ii, No. 14.
56. *Letters*, i, No. 293.
57. *Letters*, i, No. 278, n.
58. *DWJ* i. 151.
59. *Letters*, i, No. 282.
60. *DWJ* i. 415.
61. *DWJ* i. 422.
62. *Letters*, i, No. 292.
63. *Letters*, ii, No. 6.
64. *Letters*, ii, No. 31.
65. *Letters*, ii, No. 6.
66. *STC Letters*, No. 617.
67. Richard Holmes, *Coleridge*, pp. 23–5.
68. *Letters*, ii, No. 48.

CHAPTER 12

1. *Letters*, ii, No. 55.
2. *STC Letters*, No. 1062.
3. *Letters*, ii, No. 23.

4. John Nichols, *History and Antiquities of the County of Leicester* (1804), vol. iii, part II, pp. 733–44.

5. P. A. Tomory, *Sir George Beaumont and his Circle*, Catalogue to the Bicentenary Exhibition, Leicester Museum and Art Gallery (1953), Introduction.

6. *STC Letters*, No. 508.

7. *STC Letters*, No. 558.

8. *Letters*, i, No. 213.

9. *Letters*, ii, No. 55.

10. *Letters*, i, No. 239.

11. *Letters*, i, No. 282.

12. *DWJ* i. xiv.

13. *Letters*, i, No. 257.

14. *Letters*, i, No. 280.

15. *Letters*, ii, No. 51.

16. *Letters*, ii, No. 48.

17. *Letters*, ii, No. 55.

18. *Letters*, ii, No. 62.

19. *Letters*, ii, No. 57.

20. *Letters*, ii, No. 60.

21. *Letters*, ii, No. 67.

22. *Letters*, ii, No. 67.

23. S. T. Coleridge, *The Statesman's Manual: a Lay Sermon* (1816).

24. W. Wilberforce, *Practical Review of the Prevailing Religious System*, 8th edn. (1805.)

25. Southey, *Letters*, No. 138.

26. *Letters*, ii, No. 81.

27. *Letters*, ii, No. 225.

28. *Letters*, ii, No. 223.

29. *Letters*, i, No. 258.

30. *Letters*, ii, No. 33.

31. *Letters*, ii, No. 87.

32. *Letters*, ii, No. 95.

33. *Letters*, ii, No. 92.

34. *CN* iii, No. 3304.

35. *Letters*, ii, No. 91.

36. *Letters*, ii, Nos. 82 and 94.

37. *Letters*, ii, No. 120.

38. *Letters*, ii, Nos. 83 and 86.

39. *Letters*, ii, No. 101.

40. De Selincourt, p. 230.

41. *DWJ* i. 59.

42. *Letters*, i, No. 284.

43. *Letters*, i, No. 243.

44. *Letters*, ii, No. 102.

45. *Letters*, ii, No. 181.

46. *Letters*, ii, No. 211.

47. *Letters*, ii, No. 112.

48. *Letters*, ii, No. 117.

49. *Letters,* ii, No. 122.
50. *Letters,* ii, No. 99.
51. *Letters,* ii, Nos. 167 and 174.
52. *Letters,* ii, No. 135.
53. *LSH,* No. 11.
54. R. J. White (ed.), *Political Tracts of Wordsworth, Coleridge and Shelley.*
55. *LSH,* No. 11.
56. *Letters,* ii, No. 135.
57. *Letters,* ii, No. 134.
58. *Letters,* ii, No. 151.
59. *Letters,* ii, No. 158.
60. *Letters,* ii, No. 174.
61. *Letters,* ii, No. 140.
62. *Letters,* ii, No. 169.
63. *Letters,* ii, No. 158.
64. *Letters,* ii, No. 177.

CHAPTER 13

1. *Letters,* ii, No. 188.
2. *STC Letters,* No. 809, editorial note.
3. *Letters,* ii, No. 223.
4. *Letters,* iii, No. 243.
5. *Letters,* ii, No. 93.
6. *Letters,* i, No. 243.
7. *Letters,* ii, No. 176.
8. *Letters,* i, Nos. 225 and 282.
9. *Letters,* ii, No. 87.
10. *Letters,* ii, No. 158.
11. *Letters,* ii, No. 150.
12. *Letters,* ii, No. 211.
13. *LSH,* No. 77.
14. *Letters,* ii, No. 148.
15. *Letters,* ii, Nos. 143, 154, 169.
16. *Letters,* ii, No. 184.
17. *Letters,* ii, No. 205.
18. *Letters,* ii, No. 188.
19. *Letters,* ii, No. 234.
20. *Letters,* ii, No. 193.
21. *Letters,* ii, No. 192.
22. *Letters W and MW,* p. 51.
23. *Letters,* ii, No. 196.
24. *Letters,* ii, No. 206.
25. *Letters W and MW,* p. 35.
26. *Letters,* ii, No. 200.
27. *Letters,* ii, No. 203.
28. *Letters,* ii, No. 204.
29. *Letters,* ii, No. 203.

30. *Letters*, ii, No. 206.
31. *Letters W and MW*, p. 53.
32. *Letters*, ii, No. 211.
33. *Letters*, ii, No. 210.
34. *C and ML Letters* ii. 107.
35. *C and ML Letters* ii. 104.
36. *Letters*, ii, No. 208.
37. *LSH*, No. 61.
38. *Letters*, ii, No. 208.
39. *Letters*, ii, No. 209.
40. *Letters*, ii, No. 217.
41. *Letters*, ii, No. 211.
42. *Letters*, ii, No. 211.
43. *Letters*, ii, No. 214.
44. *Letters*, ii, No. 223.
45. *Letters*, ii, No. 225.
46. *LSH*, No. 12.
47. *LSH*, No. 13.
48. *Letters*, ii, No. 234.
49. *Letters*, iii, No. 240.
50. *Letters*, iii, No. 240.
51. *Letters W and MW*, p. 115.
52. *Letters W and MW*, pp. 237–9.
53. *Letters W and MW*, p. 105.
54. *Letters*, iii, No. 253.
55. *Letters*, iii, No. 253.
56. Lindop, pp. 198–200.
57. *Letters*, iii, No. 253.
58. Ibid.
59. *Letters*, iii, No. 254.
60. *Letters*, iii, No. 259.
61. *Letters*, iii, No. 273.
62. *Letters*, iii, No. 272.
63. *Letters*, iii, No. 275.
64. *Letters*, iii, No. 277.
65. *Letters*, iii, Nos. 277 and 279.

CHAPTER 14

1. *Letters*, iii, Nos. 268, 271, 274, 282, and 288.
2. *Letters*, iii, No. 292.
3. *Letters*, iii, No. 287.
4. *W Circ.*, i. 76.
5. *Letters*, iii, No. 457.
6. *Letters*, iii, No. 292.
7. *LSH*, No. 18.
8. *Letters*, iii, No. 307.
9. *Letters*, iii, No. 304.

10. *Letters*, iii, No. 305.
11. *Letters*, iii, Nos. 335a and 336.
12. *Letters*, iii, No. 369.
13. *Letters*, iii, No. 444.
14. *Letters*, iii, No. 341.
15. *Letters*, i, Nos. 278 and 279.
16. *Letters*, ii, Nos. 138 and 181.
17. *Letters*, iii, No. 378.
18. *Letters*, iii, No. 398.
19. *Letters*, i, No. 140.
20. *Letters*, i, No. 251.
21. *Letters*, ii, No. 234.
22. *Letters*, ii, No. 204.
23. *Letters*, ii, No. 223.
24. *Letters*, iii, No. 374.
25. *Letters*, iii, No. 379.
26. *Letters*, iv, No. 18.
27. Moorman, ii. 255–6.
28. *Letters*, i, No. 257.
29. *Letters*, iii, No. 448.
30. *Letters*, iii, No. 311.
31. *Letters*, iii, No. 318.
32. *Letters*, iii, No. 332.
33. *Letters*, iii, No. 350.
34. *Letters*, iii, No. 341.
35. *Letters*, iii, No. 370.
36. *Letters*, ii, No. 81.
37. *Letters*, iii, No. 350.
38. *Letters*, iii, No. 237.
39. *Letters*, iii, No. 308.
40. *Letters*, iii, No. 370.
41. *Letters*, iii, No. 280.
42. *DWJ* i. 123.
43. *Letters*, iii, No. 444.
44. Ibid.
45. *Letters*, iii, No. 318.
46. *Letters*, iii, No. 243.
47. *Letters*, iii, No. 330.
48. Ibid.
49. *Letters*, iii, No. 357.
50. *Letters*, iii, No. 359.
51. *Letters*, iii, No. 357.
52. *Letters*, iii, No. 359.
53. *Letters*, iii, No. 370.
54. Southey, Letters, No. 108.
55. Ibid. No. 109.
56. *Letters*, iii, No. 359.
57. *Letters*, iii, No. 370.

58. *Letters*, iii, No. 370.
59. The processional flights of steps described by Legouis were not constructed until the rebuilding in 1824.
60. Legouis, pp. 103–5.
61. *Letters*, iii, No. 395.
62. Legouis, pp. 106–7.
63. *Letters*, iii, No. 460.
64. *Letters*, iii, No. 433.
65. *New letters of Robert Southey*, ed. K. Curry, ii. 159–63.
66. *Letters*, iii, No. 460.
67. Legouis, pp. 106–7. This likeness was most marked in middle age.

CHAPTER 15

1. *Letters W and MW*, p. 251.
2. *Letters*, iii, No. 488.
3. *LSH*, No. 44.
4. *Letters*, iii, No. 481.
5. *Letters*, iii, No. 486.
6. *Letters*, iii, No. 486.
7. R. J. White, *Waterloo to Peterloo*, p. 38.
8. *Letters*, iv, No. 240, n. 4.
9. *Letters*, iii, No. 506.
10. *Letters*, iii, No. 467.
11. *Letters of John Keats*, i. 299.
12. *KC* ii. 61.
13. *Letters of John Keats*, i. 302–3.
14. Southey, *Letters*, No. 135.
15. *Letters*, iii, No. 506.
16. *Letters*, iii, No. 514.
17. *Letters*, iii, No. 337.
18. *Letters*, iii, No. 518.
19. S. Potter (ed.), *Minnow among Tritons*, p. 76.
20. *Letters*, ii, No. 223.
21. *LSH*, No. 14.
22. *Letters*, iii, No. 530.
23. S. Potter (ed.), *Minnow among Tritons*, pp. 77–8.
24. *Letters*, iii, No. 586.
25. *Letters*, iii, No. 552.
26. *HCR* i. 188.
27. *Letters*, iii, No. 514 and 537.
28. *Letters*, iv, No. 107.
29. *Letters*, iii, No. 565.
30. *Letters*, iii, No. 527.
31. *Rules and Orders relating to Charterhouse and to the Good Government thereof* (1748).
32. *Letters*, iii, No. 524.
33. *Letters*, iii, No. 530.

34. *LSH*, No. 56; *Letters C and ML*, ii, 265.
35. *Letters*, iii, No. 554.
36. *Letters*, iii, No. 537.
37. *Letters*, iii, No. 586.
38. J. Woodforde, *False Teeth*, p. 49.
39. *Letters C and ML*, ii. 277–8.
40. Thomas Shepherd, *London and its Environs* (1829), vol. i, plate 82 and pp. 77–8.
41. *Letters*, iii, No. 586.
42. *Letters*, iii, No. 589.
43. *Letters*, iii, No. 586.
44. *Letters*, iii, No. 592.
45. *LSH*, No. 64.
46. *Letters*, iii, No. 596.
47. Mary Wordsworth's Journal, WL MS.
48. *Letters*, iv, No. 46.
49. *Letters*, iv, No. 139, n. 1.
50. *Letters*, iii, No. 600.
51. DC 92, fol. 163–4, WL.
52. DC 90, vol. ii, fol. 325, WL.
53. DC 92, back flyleaf, WL.
54. DC 90, fol. 348–9, WL.
55. *W Circ.* i. 101-2.
56. Legouis, p. 109.
57. *Night and Day*, 11 November 1937, p. 27.

CHAPTER 16

1. K. Curry (ed.), *New Letters of Robert Southey*, ii. 284–5.
2. *HCR* i. 296–7.
3. *Letters*, iii, No. 47; v, Nos. 526 and 588.
4. White, *History, Gazetteer and Directory of Suffolk.*
5. *W Circ.* i. 159.
6. *Letters*, iv, No. 8.
7. D. Winstanley, *Early Victorian Cambridge*, pp. 61–79.
8. *Letters*, iv, Nos. 8 and 19.
9. Andrew Bell, *Elements of Tuition.*
10. *Letters*, iv, No. 30.
11. *Letters*, iv, Nos. 69 and 70.
12. *Letters*, iv, No. 62 and n.
13. *Letters*, iv, No. 52.
14. *Letters*, iv, No. 46.
15. *LSH* No. 97.
16. *Letters*, iv, No. 70.
17. *LSH*, No. 241.
18. *Letters*, iv, No. 153.
19. *Letters*, iv, No. 220.
20. *Letters*, iv, No. 56.

21. *Letters*, iv, No. 83.
22. *Letters*, iv, Nos. 74 and 105.
23. *Letters*, i, No. 192 and ii, No. 119.
24. *DWJ* ii. 394.
25. *Letters*, iv, Nos. 99 and 102.
26. *Letters*, iv, Nos. 104 and 108.
27. *Letters*, iv, No. 116.
28. *Letters*, iv, No. 116.
29. *W. Circ.* i, 195.
30. *Letters*, iv, No. 132.
31. *Letters*, iv, No. 139.
32. *Letters*, iii, No. 565.
33. *Letters*, iv, No. 139.
34. *Letters*, iv, No. 166.
35. *LSH*, No. 310, and *Letters*, iv, No. 195.
36. Lindop, p. 279.
37. *Letters*, iv, No. 198.
38. *LSH*, No. 118.
39. DC 104, WL.
40. *Letters*, iv, No. 221.
41. *LSH*, No. 116.
42. DC 104, WL.
43. DC 104, 10–16 September 1826, WL.
44. DC 104, 2–30 October 1826, WL.
45. *Letters*, iv, No. 260.
46. *Letters*, iv, No. 272.
47. DC 104, WL.
48. Ibid.
49. Ibid.
50. *Letters*, iv, No. 317.
51. DC 104, WL.
52. *LSH*, No. 133.

CHAPTER 17

1. *LSH*, No. 77.
2. *Letters*, iv, No. 86.
3. *Letters*, iv, No. 159.
4. *Letters*, iv, No. 203.
5. *Letters*, iv, No. 266.
6. *Letters*, iv, No. 279.
7. *Alumni Cantabrigiensis*.
8. *Letters*, iv, No. 293.
9. *Letters*, iv, No. 331.
10. *Letters*, iv, No. 337.
11. *Letters*, iv, No. 365.
12. DC 104, WL.
13. *Letters*, iv, p. 666, n.1.

14. *Letters*, iv, No. 392.
15. *Letters*, i, No. 28.
16. DC 104, accounts entered in reverse at end pages of 1828 journal, WL.
17. *Letters*, iv, No. 379.
18. John Nichols, *History and Antiquities of the County of Leicester* (1804), vol. iii, part II, pp. 112–22.
19. *Letters*, iv, No. 379.
20. *Letters*, iv, No. 377.
21. DC 104, WL.
22. *W Circ.* i. 203.
23. DC 104, WL.
24. *Letters*, v, Nos. 424 and 425.
25. *Letters*, v, No. 428.
26. WL MS A/105.
27. *Letters*, v, No. 463.
28. DC 104, WL.
29. *Letters*, v, No. 494.
30. *Letters*, v, No. 463.
31. *Letters*, v, No. 471.
32. *Letters*, v, No. 476.
33. *Letters*, v, No. 481.
34. *Letters*, v, No. 492.
35. Guildhall MS 8241/3.
36. T. Graham, *Domestic Medicine*, 6th ed., p. 258.
37. *Letters*, v, No. 494.
38. *Letters*, i, No. 164.
39. W. S. Hanbrich in *Gastroenterology*, vol. ii, pp. 895–903.
40. *Letters*, v, No. 494.
41. *Letters*, v, No. 522.
42. DC 118, WL.
43. DC 118, WL.
44. *Letters*, v, No. 574, 576.
45. *Letters*, v, No. 590.
46. DC 118, WL.
47. DC 118, WL.
48. *Letters*, v, No. 639.
49. *LSH*, No. 147.
50. *Letters*, v, No. 710.
51. *Letters*, v, No. 711.
52. *DWJ* i. 440, App. 2.
53. *Letters*, v, No. 727.
54. *Letters*, v, No. 730.
55. *Letters*, v, No. 741.
56. For the standard dosage see W. Buchan, *Domestic Medicine*, 21st edn., pp. 294–7, and T. Graham, *Domestic Medicine*, 6th edn., p. 506. Opium was prescribed for this condition into the twentieth century.
57. *Letters*, vi, No. 1163.
58. DC 118, WL.

59. De Selincourt, p. 390. Journal pages for this time removed since 1933.
60. *Letters*, v, No. 796.
61. F. Blanshard, *Portraits of Wordsworth*, p. 160.
62. DC 118, WL.
63. *Letters*, v, No. 805.
64. *Letters*, v, No. 815.
65. DC 118, WL.
66. Coleridge, *Table Talk*, 27 September 1830.
67. *STC Letters*, vi, No. 1788.
68. *LSH*, No. 158.
69. *LSH*, No. 159.
70. *Letters*, v, No. 831.
71. *Letters*, v, No. 843.
72. This difference of interpretation has been continued by twentieth-century writers: e.g. E. Batho, *The Later Wordsworth*, pp. 390–4 and E. Legouis , *Revue Anglo-Americaine*, Oct. 1923, pp. 66–73.
73. WL MSS. See Appendix One.

CHAPTER 18

1. *Letters*, vi, No. 866.
2. DC 118, WL.
3. *Letters*, vi, No. 874.
4. *Letters*, vi, No. 885.
5. *Letters*, vi, No. 892.
6. *Letters*, vi, No. 899.
7. *Letters*, vi, No. 928.
8. *Letters*, vi, No. 929.
9. *Letters*, vi, No. 928.
10. *Letters*, vi, No. 941.
11. Information, Alan G. Hill.
12. *W Circ*. i. 282.
13. G. E. W. Wolstenholme and M. Binney (ed.), *Alzheimer's Disease and Related Conditions*. See Appendix Two.
14. *W Circ*. i. 287–8.
15. *LMW*, Nos. 78 and 81.
16. Information, Alan G. Hill.
17. *LMW*, No. 84.
18. Information kindly supplied by Roger Fiske.
19. *W Circ*. i. 271.
20. *LMW*, No. 76.
21. *HCR* iii. 78.
22. DC 118, WL.
23. *Letters*, vi, No. 929.
24. *LMW*, No. 72.
25. *Letters*, vi, No. 989.
26. *Letters*, vi, No. 941.
27. *LMW*, No. 76.

28. *HCR* iii. 78.
29. *Letters*, vi, No. 1176.
30. *Letters*, vi, No. 1207.
31. *Letters*, vi, No. 1056.
32. *W Circ.* i. 421.
33. Julia Wedgwood, *The Personal Life of Josiah Wedgwood the Potter*, p. 332.
34. *W Circ.* ii. 725.
35. Moorman, ii. 607.
36. WL MS. Hitherto unpublished. Information, Alan G. Hill.
37. *LMW*, Nos. 177 and 178; cf. De Selincourt, 416–7.

Bibliography

Alumni Cantabrigiensis.
Alumni Oxoniensis.

Barton, Mary E. (ed.), *Letters of Mary Wordsworth*, 1958.
Batho, E., *The Later Wordsworth*, 1933.
Bayne-Powell, Rosamond, *Housekeeping in the Eighteenth Century*, 1956.
Bell, Andrew, *Elements of Tuition*, 1797.
Blanshard, Frances, *Portraits of Wordsworth*, 1959.
Blomefield, *Topographical History of Norfolk*, 1806.
Blunden, Edmund, and Griggs, Earl Leslie (eds.), *Coleridge: Essays by Several Hands*, 1934.
British Medical Journal.
Broughton, L. N. (ed.), *Some Letters of the Wordworth Family*, 1942.
Buchan, W., *Domestic Medicine*, 21st ed., 1813.
Campbell, J. Dykes, *Samuel Taylor Coleridge*, 1894.
Charterhouse, Governors of, *Rules and Orders relating to Charterhouse* etc., 1748.
Coburn, Kathleen (ed.), *The Letters of Sara Hutchinson from 1800 to 1835*, 1954.
    The Notebooks of Samuel Taylor Coleridge, 3 vols., 1957-.
Coleridge, S. T. C., *Biographia Literaria*, 1817.
    Table Talk of S. T. Coleridge, 1835.
Cottle, J., *Reminiscences of Coleridge and Southey*, 1848.
Curry, K. (ed.), *New Letters of Robert Southey*, 2 vols., 1965.
Darlington, Beth (ed.), *The Love Letters of William Mary Wordsworth*, 1982.
Darby, P., *An Essay on Halifax*, 1761.
De Selincourt, Ernest, *Dorothy Wordsworth*, 1933.
De Selincourt, E. (ed.), *Journals of Dorothy Wordsworth*, 2 vols., 1941.
De Selincourt, E. and Darbyshire, H. (eds.), *The Poetical Works of William Wordsworth*, 5 vols., 1940-.
Dyce, A., *Recollections of the Table Talk of Samuel Rogers*, 1887.
Eccles, R., *Northgate End Chapel, Halifax*, 1946.
Eden, Sir Frederick Morton, *The State of the Poor*, [1795], abridged and edited A. G. L. Rogers, 1928.
Fitzgerald, M. H. (ed.), *The Letters of Robert Southey, a Selection*, 1912.
Gentleman's Magazine.
Glasse, H., ("A Lady"), *The Art of Cookery made Plain and Easy*, 21st ed., 1796.
Graham, Thomas J., *Modern Domestic Medicine*, 6th edition, 1835.
Greaves, Margaret, *Regency Patron: Sir George Beaumont*, 1966.
Griggs, Earl Leslie, *Thomas Clarkson, the Friend of Slaves*, 1936.
    (ed.) *Collected Letters of Samuel Taylor Coleridge*, 6 vols., 1956-1971.
Hill, Alan G. (general editor), *Letters of William and Dorothy Wordsworth*, 1967-.
Hill, Alan G., in *Review of English Studies*, XXXI, No, 123.
    "*Wordsworth and the Two Faces of Machiavelli*".
Holmes, R., *Coleridge*, 1983.
House, Humphry, *All in Due Time*, 1955.
Hutchinson, W., *The History and Antiquities of Cumberland*, 1794.

Jordan, J. E., *De Quincey to Wordsworth*, 1962.

Keats, John, *Letters*, ed. H. E. Rollins, 2 vols., 1958.

Knight, W. (ed.), *Journals of Dorothy Wordsworth*, 1897.

Lawrence, Berta, *Coleridge and Wordsworth in Somerset*, 1970.

Legouis, Emile, *William Wordsworth and Annette Vallon*, 1922.

Lindop, Grevel, *The Opium Eater: a Life of Thomas De Quincey*, 1982.

Lister, Anne, *Diary of a Halifax Lady*, repr. in *The Halifax Guardian*, 1887-1889.

Lucas, E. V. (ed.), *The Letters of Charles and Mary Lamb*, 3 vols., 1935.

Maclean, C., *Dorothy Wordsworth: The Early Years*, 1932.

Millison, F. E., *Two Hundred Years of the Northgate End Chapel*, 1896.

Moorman, M., *William Wordsworth: a Biography*, 2 vols., 1957 and 1965.

Morley, E. J. (ed.), *Correspondence of Henry Crabb Robinson with the Wordsworth Circle*, 1935.

    Henry Crabb Robinson on Books and Their Writers, 1938.

Nichols, J., *History and Antiquities of the County of Leicester*, 1804.

Oxford History of English Literature.

Pennington, M., *Memoirs of the Life of Mrs. Elizabeth Carter*, 1807.

Population Census, 1801.

Potter, S. (ed.), *Minnow among Tritons*, 1934.

Quillinan, E., *Poems*, 1851.

Quincey, T. De, *Recollections of the Lakes and the Lake Poets*, ed. D. Wright, 1970.

Rand, F. P., *Wordsworth's Mariner Brother*, 1966.

Reed, Mark L. (ed.), *Wordsworth: the Chronology of the Early Years*, 1967.

Sandford, H., *Thomas Poole and His Friends*, 1888.

Shepherd, Thomas, *London and its Environs*, 1829.

Stenton, D. M., *The English Woman in History*, 1957.

Thomas, K., *Man and the Natural World*, 1983.

Thompson, J. M., *The French Revolution*, 1944.

Tomalin, C., The life and Death of Mary Wollstonecraft, 1974.

Tomory, P. A., *Sir George Beaumont and his Circle*, 1953.

Trimmer, S., *The Oeconomy of Charity*, 1786.

Wedgwood, J., *The Personal Life of Josiah Wedgwood the Potter*, ed. Farrer, 1906.

White, ——, *History, Gazeteer, and Directory of Suffolk*, 1840.

White, R. J. (ed.), *Political Tracts of Wordsworth, Coleridge, and Shelley*, 1955.
    Waterloo to Peterloo, 1957.

Wilberforce, W., *Practical Review of the Prevailing Religious System*, 8th ed., 1805.

Winstanley, D. S., *Early Victorian Cambridge*, 1955.

Wolstenholme, G. E. W. and Binney, M. (eds.), *Alzheimer's Disease and Related Conditions*, Ciba Foundation Symposium, 1970.

Woodforde, James, *Diary*, 1924-1931.

Woodforde, John, *False Teeth*, 1968.

Index